Berichte aus dem Institut für Mehrphasenströmungen

Band 10

Hydrodynamic Characterization of Heterogeneities in Aerated Stirred Tank Reactors

From an Eulerian to a Lagrangian Perspective

Vom Promotionsausschuss der
Technischen Universität Hamburg
zur Erlangung des akademischen Grades
Doktor-Ingenieur (Dr.-Ing.)

genehmigte Dissertation

von
Jürgen Fitschen

aus
Zeven

2022

Bibliografische Information der Deutschen Nationalbibliothek
Die Deutsche Nationalbibliothek verzeichnet diese Publikation in der
Deutschen Nationalbibliografie; detaillierte bibliographische Daten sind im Internet
über http://dnb.d-nb.de abrufbar.
1. Aufl. - Göttingen: Cuvillier, 2022
Zugl.: (TU) Hamburg, Univ., Diss., 2022

1. Gutachter: Prof. Dr.-Ing. habil. Michael Schlüter
2. Gutachter: Prof. Dr. Miroslav Šoóš
3. Gutachter: Dr.-Ing. Thomas Wucherpfennig

Vorsitzender: Prof. Dr.-Ing. Dr. h.c. Stefan Heinrich

Tag der mündlichen Prüfung: 18.02.2022

Nonnenstieg 8, 37075 Göttingen
Telefon: 0551-54724-0
Telefax: 0551-54724-21
www.cuvillier.de

1. Auflage, 2022
Gedruckt auf umweltfreundlichem, säurefreiem Papier aus nachhaltiger Forstwirtschaft.

ISBN 978-3-7369-7689-4
eISBN 978-3-7369-6689-5

Foreword

This dissertation is the result of my work as a research associate at the *Institute of Multiphase Flows* of the *Hamburg University of Technology* under the guidance of my doctoral advisor Prof. Dr.-Ing. habil. Michael Schülter, whom I sincerely thank for his trust, professional expertise and tireless support in the preparation of the thesis. I want to thank Prof. Dr. Miroslav Soos for his fascinating, interdisciplinary collaboration and his readiness to evaluate my work as a second examiner. I also wish to thank Prof. Dr.-Ing. Dr. h.c. Stefan Heinrich for taking on the role of chair of my dissertation committee.

Additional sincere thanks are owed to Dr.-Ing. Thomas Wucherpfennig from our long-standing partnership with Boehringer Ingelheim Pharma GmbH Co. KG for his confidence and technical contribution as an external committee member for my work. The experiences and scientific insights which I was privileged to gather in the joint cooperation contributed in large part to the success of this work. At this point special thanks are owed to my doctoral advisor, who placed his trust in me and gave me the chance to take part in this industrial cooperation. I always felt that the opportunity to regularly present our work and results at national and international conferences was also very special.

Furthermore, I owe special thanks to Prof. Dr. Alexandra von Kameke, not only for her expert evaluation and valuable comments on my work, but especially for her tireless and enduring assistance in carrying out the 4DPTV measurements in the laboratory. Despite adverse conditions due to the corona restrictions at the time, Alex did not let herself get discouraged and was still in the laboratory with me for hours on end performing and optimizing the measurements until shortly before Christmas 2020.

Additionally, I want to thank Dr.-Ing. Johannes Wutz for his collaboration in the industrial cooperation with respect to the mixing time evaluations and the further development of the methodology. Our joint night shifts solving problems which cropped up while analyzing experimental data will especially remain in my memory forever. Likewise, I want to send sincere thanks to Dr.-Ing. Maike Kuschel, who during several professional and personal discussions gave me deep insights into fascinating, actual pharmaceutical upstream processes. All these insights, experiences and collaborations have contributed immeasurably to the success of my work.

Furthermore, I wish to sincerely thank all my colleagues for the wonderful teamwork in the laboratory and classroom, as well as for the great business trips. Special thanks go to my predecessor Dr.-Ing. Annika Rosseburg, from whom I learned a great deal about practical work as a scientist and engineer. Many thanks, dear Annika, for your support. Special heartfelt thanks go to Dr.-Ing. Marko Hoffmann who, besides his technical expertise, always made possible every request regarding measuring equipment as well as participation in conferences, in particular international ones. In addition I would like to sincerely thank my office mates from those times, Christian Busch and Dr.-Ing. Simon Matthes, as well as Felix Kexel, who is still my office mate, for the wonderful countless scientific discussions, as well as our personal discussions, out of which a genuine friendship developed.

Likewise, I wish to sincerely thank my former students and current colleagues Ingrid Haase and Vincent Bernemann, who not only turned in outstanding assignments, but also allowed me to play a part in their development from university students to doctoral candidates. I also want to warmly thank Mr. Bernhard Pallaks for his immeasurably valuable assistance and suggestions for the implementation of my innumerable ideas in the laboratory and workshop, which spared me a great deal of time and nerves. Likewise, I would like to sincerely thank my undergraduate assistants Helena Ostrinsky and Jakob Schulze.

Finally, I would like to express my sincere thanks to my family and friends for always supporting and encouraging me on my path to my doctorate. Very special thanks go to my parents Bettina and Klaus Fitschen, whose love and support has left its imprint on my entire scientific education and made it possible. Last but not least I owe thanks to my companion Lara-Christin, who not only supported me emotionally but always had my back and always stands by my side.

Hamburg, February 2022

Vorwort

Die vorliegende Arbeit entstand während meiner Zeit als wissenschaftlicher Mitarbeiter am *Institut für Mehrphasenströmungen* an der *Technischen Universität Hamburg* unter der Leitung meines Doktorvaters Prof. Dr.-Ing. habil. Michael Schlüter, welchem ich herzlich für sein Vertrauen, seine fachliche Expertise und unermüdliche Hilfestellung bei der Erstellung der Thesis danke. Bedanken möchte ich mich weiter bei Herrn Prof. Dr. Miroslav Soos für die spannende und interdisziplinäre Zusammenarbeit und die Bereitschaft, meine Arbeit als Zweitgutachter zu bewerten. Bedanken möchte ich mich auch bei Herrn Prof. Dr.-Ing. Dr. h.c. Stefan Heinrich für die Übernahme des Prüfungsvorsitzenden.

Darüber hinaus gilt mein herzlicher Dank Herrn Dr.-Ing. Thomas Wucherpfennig von unserem langjährigen Kooperationspartner Boehringer Ingelheim Pharma GmbH Co. KG für sein Vertrauen und seinen fachlichen Beitrag als externer Gutachter zu meiner Arbeit. Die Erfahrungen und wissenschaftlichen Erkenntnisse, welche ich in der gemeinsamen Kooperation sammeln durfte, haben großteilig zum Erfolg der vorliegenden Arbeit beigetragen. An dieser Stelle gilt ein besonderer Dank auch meinem Doktorvater, welcher mir sein Vertrauen schenkte und mir die Chance gab, Teil diese Industriekooperation sein zu dürfen. Auch die Möglichkeit unsere Arbeiten und Ergebnisse regelmäßig bei nationalen und internationalen Kongressen vorstellen zu dürfen, empfand ich stets als etwas ganz Besonderes.

Darüber hinaus gilt mein besonderer Dank Frau Prof. Dr. Alexandra von Kameke nicht nur für Ihre fachliche Begutachtung und wertvollen Anmerkungen meiner Arbeit, sondern insbesondere auch für die unermüdliche und ausdauernde Unterstützung bei der Durchführung der 4DPTV Messungen im Labor. Trotz den widrigen Umständen, welche uns die damaligen coronabedingten Einschränkungen bereitet haben, hat sich Alex nicht entmutigen lassen, noch bis kurz vor Weihnachten 2020 stundenlang mit mir im Labor zu sein und die Messungen zu optimieren und durchzuführen.

Des Weiteren möchte ich mich bei Herrn Dr.-Ing Johannes Wutz, für die Zusammenarbeit in der Industriekooperation im Bereich der Mischzeitcharakterisierung und der Weiterentwicklung der Methodik herzlich bedanken. Insbesondere unsere gemeinsamen Nachtschichten bei der Lösung von Problemen, welchen wir bei der Analyse der experimentalen Daten ausgesetzt waren, werden mir ewig in Erinnerung bleiben. Ebenfalls möchte ich mich herzlich bei Frau

Dr.-Ing Maike Kuschel bedanken, welche mir in vielen fachlichen Diskussionen, aber auch in persönlichen Gesprächen die Möglichkeit gegeben hat, tiefe Einblicke in hochspannende und reale pharmazeutische Upstream-Prozesse zu bekommen. All diese Einblicke, Erfahrungen und Kollaborationen zusammen haben einen unmessbar wertvollen Beitrag zum Erfolg dieser Arbeit beigetragen.

Weiterhin möchte ich mich herzlich bei all meinen Kollegen:innen für die tolle Zusammenarbeit im Labor und der Lehre, aber auch den tollen gemeinsamen Dienstreisen bedanken. Ein besonderer Dank gilt zudem meiner Vorgängerin Dr.-Ing. Annika Rosseburg, von welcher ich viel über die Arbeit als Wissenschaftler und Ingenieur in der Praxis gelernt habe. Vielen lieben Dank, liebe Annika, für deine Unterstützung. Ein besonders herzlicher Dank gilt Herrn Dr.-Ing. Marko Hoffmann, welcher neben seiner fachlichen Expertise stets im Hintergrund sämtliche Wünsche hinsichtlich neuen Messequipments, aber auch die Teilnahmen an insbesondere internationalen Kongressen mit ermöglicht hat. Darüber hinaus möchte ich mich herzlich bei meinen damaligen Bürokollegen Christian Busch und Dr.-Ing. Simon Matthes und immer noch Kollegen Felix Kexel für die tollen und unzähligen fachlichen Diskussionen, aber auch für die persönlichen Gespräche bedanken, aus welchen sich in den Jahren eine echte Freundschaft entwickelt hat. Gleichermaßen möchte ich mich bei meinen ehemaligen Studierende und jetzigen Kollegen Ingrid Haase und Vincent Bernemann herzlich bedanken, welche nicht nur hervorragende studentische Arbeiten vorzuweisen haben, sondern mich auch bei Ihrer persönlichen Entwicklung vom Studierenden bis zum Doktoranden haben mitwirken lassen. Überdies möchte ich mich herzlich bei Herrn Bernhard Pallaks für seine unmessbar wertvollen Hilfestellungen und Ratschläge zur Umsetzung meiner unzähligen Ideen im Labor und der Werkstatt bedanken, welche mir viel Zeit und Nerven bei der Umsetzung erspart haben. Ebenfalls möchte ich mich herzlich bei meinen studentischen Hilfskräften Helena Ostrinsky und Jakob Schulze bedanken. Letztlich möchte ich meiner Familie und meinen Freunden herzlich Danke sagen dafür, dass Sie mich stets auf dem Weg zur Promotion unterstützt und ermutigt haben. Ein ganz besonderer Dank gilt jedoch meinen Eltern Bettina und Klaus Fitschen, deren Liebe und Unterstützung meine gesamte wissenschaftliche Ausbildung ermöglicht und geprägt hat. Last but not least geht mein liebster Dank an meine Freundin Lara-Christin, welche mich nicht nur mental unterstützt, sondern mir stets den Rücken freigehalten hat und immer an meiner Seite steht.

Hamburg, im Februar 2022

Für meine Eltern, Bettina & Klaus.

Contents

List of Figures

List of Tables

Nomenclature

Roman Symbols

A	s	exposure time
$\boldsymbol{a}$	$\mathrm{m \cdot s^{-2}}$	acceleration
C_k		Kolmogorov constant ≈ 1.5
d_stirrer	m	stirrer diameter
d_p	m	particle diameter
D_STR	m	tank diameter
D_1	$\mathrm{m^2 \cdot s^{-2}}$	ballistic dispersion coefficient
D_2	$\mathrm{m^2 \cdot s^{-1}}$	diffusive dispersion coefficient
E	$\mathrm{m^2 \cdot s^{-2}}$	absolute specific kinetic energy
f	s	exposure time
H	m	tank height
h	m	bottom clearance
I		gray value
k	$\mathrm{m^{-1}}$	wave number
L	m	length
M	$\mathrm{N \cdot m}$	torque
n	$\mathrm{s^{-1}}$	stirrer frequency
P	W	power

q	$m^3 \cdot s^{-1}$	gas volume flow
r	m	displacement
r	m	radius
s	m	stirrer spacing
t	s	time
T_L	s	Lagrangian integration time
T	K	temperature
u	$m \cdot s^{-1}$	velocity, x direction
v	$m \cdot s^{-1}$	velocity, y direction
$\boldsymbol{v}$	$m \cdot s^{-1}$	velocity vector
V	m^3	volume
w	$m \cdot s^{-1}$	velocity, z direction
X		blending ratio

Dimensionless numbers

$Re_{\text{stirrer}} = \frac{nd^2\rho}{\eta}$	Stirrer Reynolds number
$Fr = \frac{n^2}{gd}$	Froude number
$Po = \frac{P}{\rho n^3 d^5}$	power number

Greek Symbols

λ		dimensionless vertical direction
θ		dimensionless mixing time
μ		dimensionless scaling factor
ρ	$kg \cdot m^{-3}$	density
η	$kg \cdot m^{-1} \cdot s^{-1}$	dynamic viscosity

ν	$m^2 \cdot s^{-1}$	kinematic viscosity
Π		any dimensionless number
Θ	s	mixing time
ε	$W \cdot m^{-3}$	energy dissipation rate
σ		variance
F		Lagrangian auto correlation function
Ψ		distribution function
τ		distribution time
κ		proportionality coefficient

Subscripts

tip	stirrer tip
stirrer	stirrer
1	status 1
2	status 2
total	total
bearing	bearing
k	wave number
ν	viscosity range
norm	normalized
avg	average
fill	reactor filling volume
A	position of acid addition

Nomenclature

Acronyms / Abbreviations

LED	light emitting diode
STR	stirred tank reactor
LIF	laser induced fluorescence
MIG	multi-stage pulse counter-current stirrer
PTU	programmable timing unit
EE	Elephant Ear turbine
4D	four-dimensional
4DPTV	4D particle tracking velocimetry
PIV	particle image velovimetry
RT	Rushton turbine

Abstract

Aerated stirred tank reactors are widely used in process engineering for chemical, biological and pharmaceutical processes due to their robust design and diverse applications. Moreover, the characterization of aerated stirred tank reactors has already been extensively described in the respective literature. However, there are still challenges, for example, in the scale-up of aerated stirred tank reactors. Especially for highly specialized processes, such as the cultivation of mammalian cells for the production of glycoproteins for therapeutic applications, the right growth conditions are required at all reactor scales to ensure maximum and consistent product quality. This can take years of scale-up process development to find the optimal operating parameters for each cell line at each reactor scale.

In the present thesis, high spatial and temporal resolution characterization experiments were performed to describe heterogeneities in a 3 L stirred tank reactor with different stirrer geometries. A double Rushton as well as a triple Elephant Ear stirrer configuration were used for the experiments. Here, the characterization experiments are based on the investigation of the specific stirrer power input, the global and local mixing time as well as the investigation of instantaneous three-dimensional flow fields using Lagrangian particle tracking. For this purpose, Lagrangian trajectories of buoyancy-neutral particles were experimentally measured throughout the reactor volume for the first time using the "Shake-The-Box" method. This enables the application of established Lagrangian analysis methods for mixing and mass transfer processes in stirred tank reactors.

The specific mechanical power input characteristics were successfully compared with literature data to validate the experimental setup. Based on the measured power input characteristics, the instantaneous velocity fields could also be experimentally validated. Furthermore, using the method of local mixing time distribution, the history of mixing has been determined and described. Based on the local mixing time, heterogeneities in the mixing of the reactor volume were identified and their influence on the global mixing was quantified. Local mixing time distributions have been further used to visualize the breakup of coherent structures.

A new approach for equipment characterization provides the analysis of Lagrangian trajectories in terms of local residence time, exposure to external accelerating forces, and global and local absolute dispersion. The residence time and exposure distribution of the trajectories provide detailed insight into how long buoyancy-neutral particles remain in specific reactor regions and how long these particles are exposed to external acceleration forces. Finally, the single-phase mass transport in the stirred tank reactor was quantified using the Lagrangian trajectories and the absolute dispersion. Furthermore a dimensionless dispersion coefficient $\overline{D_2^*}(u_{\text{tip}}) = 0.177 \pm 0.006$ was derived empirically for two investigated reactor configurations. This dimensionless dispersion coefficient provides the basis for scale-up studies, where local transport phenomena and residence times of cell trajectories are considered in addition to the global mixing time. In this regard, the three-dimensional representation of the local dimensionless dispersion coefficients offers completely new approaches for the characterization of compartments in stirred tank reactors.

Zusammenfassung

Begaste Rührkesselreaktoren gehören aufgrund der robusten Bauweise und der umfangreichen Einsatzmöglichkeiten zu den am häufigsten verwendeten verfahrenstechnischen Apparaten in der chemischen, biologischen und pharmazeutischen Industrie. Ihre Charakterisierung ist in der einschlägigen Fachliteratur bereits sehr umfangreich beschrieben, jedoch zeigt diese auch, welche Hürden unter anderem beim Scale-up von begasten Rührkesselreaktoren noch immer bestehen. Insbesondere für hoch spezialisierte Prozesse, wie die Kultivierung von Säugetierzellen, müssen optimale Wachstumsbedingungen in allen Reaktormaßstäben gewährleistet werden, um eine maximale und konstante Produktqualität zu garantieren. Dies erfordert teilweise einen jahrelangen Prozess zur Maßstabsübertragung für jede Zelllinie in den unterschiedlichen Reaktormaßstäben.

In der vorliegenden Arbeit werden örtlich und zeitlich hoch aufgelöste Charakterisierungsexperimente zur Beschreibung von Heterogenitäten in einem begasten 3 L Rührkesselreaktor mit unterschiedlichen Rührergeometrien gezeigt. Die Charakterisierungsexperimente basieren auf der Untersuchung der spezifischen Rührerleistung, der globalen und lokalen Mischzeit sowie der Untersuchung instantaner dreidimensionaler Strömungsfelder mittels Lagranger Partikelverfolgung. Dazu wurden erstmalig Lagrange Trajektorien von auftriebsneutralen Partikeln im gesamten Reaktorvolumen mit der „Shake-The-Box“ Methode experimentell ermittelt, wodurch die Anwendung von etablierten Lagrangen Analysemethoden auf Vermischungs- und Stofftransportprozesse in Rührkesselreaktoren ermöglicht wurde.

Zur Qualifizierung des experimentellen Aufbaus wurde die spezifische mechanische Leistungseintragscharakteristik erfolgreich mit Literaturdaten verglichen. Basierend auf den gemessenen Leistungseintragskennzahlen wurden ebenfalls die instantanen Geschwindigkeitsfelder experimentell analysiert. Mit der Methode der lokalen Mischzeitbestimmung kann die Historie der Vermischung ermittelt und beschrieben werden. Auf Basis der lokalen Mischzeit wurden Heterogenitäten bei der Vermischung des Reaktorvolumens identifiziert und deren Einfluss auf die globale Vermischung quantifiziert. Weiterhin konnte mit der lokalen Mischzeitverteilung das Aufbrechen von kohärenten Strukturen visualisiert werden.

Einen neuen Ansatz zur Charakterisierung von Rührkesselreaktoren stellt die Analyse Lagranger Trajektorien dar. Hiermit lassen sich lokale Verweilzeiten, die Einwirkzeit äußerer Beschleunigungskräfte sowie die globale und lokale absolute Dispersion bestimmen. Die Verweilzeit- und Expositionsverteilung der Trajektorien geben einen detaillierten Aufschluss darüber, wie lange auftriebsneutrale Partikel in bestimmten Reaktorregionen verweilen und wie lange diese Partikel äußeren Beschleunigungskräften ausgesetzt sind. Weiterhin wurde der einphasige Stofftransport im Rührkesselreaktor anhand der Lagrangen Trajektorien und der absoluten Dispersion quantifiziert. Abschließend wurde der dimensionslose Dispersionkoeffizient $\overline{D_2^*}(u_{\mathrm{tip}}) = 0.177 \pm 0.006$ für zwei untersuchte Reaktorkonfigurationen empirisch hergeleitet. Die dimensionslose Dispersion bietet die Basis für anschließende Untersuchungen zur Maßstabsübertragung, bei der ergänzend zur globalen Mischzeit auch die lokalen Transportphänomene und Verweilzeiten der Zelltrajektorien berücksichtigt werden. Die dreidimensionale Darstellung der örtlichen Dispersionskoeffizienten bietet völlig neue Möglichkeiten zur Charakterisierung der Fluiddynamik in Rührkesselreaktoren.

1. Introduction

"Bioreactor agitator engineering is a broad mosaic. The image is simple and clear from a distance, but as the viewer moves closer, a multitude of distinct individual pieces come into view." from Keith Flanegan.

The quote from Keith Flanegan impressively summarizes not only the historical background of stirred tank reactors, but also the current state of research in just one sentence. Stirred tank reactors have been used for centuries for mixing and homogenizing several substances. Whereas the focus was initially on the preparation of food and stirring was often uncontrolled, stirred tank reactors have become indispensable in modern process engineering [Kra14].

In modern processes in the chemical and biotechnology industries, various types of bubble columns and stirred tank reactors are widely used. In general, aerated stirred tank reactors are used for many applications and especially for shear-sensitive processes, such as the cultivation of mammalian cells. The use of stirred tank reactors is mainly based on the fact that, due to the mechanical power input by the stirrer, controlled operating conditions can be maintained with regard to important properties such as good heat and mass transfer as well as good mixing. Nevertheless, new processes are always developed on a laboratory scale to save time and resources. After a successful development phase, however, the process developed at laboratory scale must be scaled up to industrial scale. Especially when scaling up processes in stirred tank reactors, a good understanding of the operating behavior depending on the operating parameters such as the mechanical power input or the aeration rate can determine the success or failure of the scale-up.

For the successful scale-up of a stirred tank reactor, the initial system must be sufficiently known and characterized, and the right scale-up strategy must be selected. In particular, the choice of the scale-up strategy poses a great challenge due to the fact that not all system properties scale equally in the geometric scaling of stirred tank reactors. This is a great challenge, especially for multiphase processes. An impressive visualization of the dilemma that arises during geometric scale-up is shown in Figure 1.1 according to Zehner 2002 [Zeh02]. Using the example of the operating parameters of the dimensionless mixing time θ, the Froude number Fr, the stirrer tip speed u_{tip} and the stirrer Reynolds number Re, Figure 1.1 shows the dependence of the specific power input P/V for different orders of magnitude where one of the operating parameters is kept constant. It can be seen, that for a constant specific power

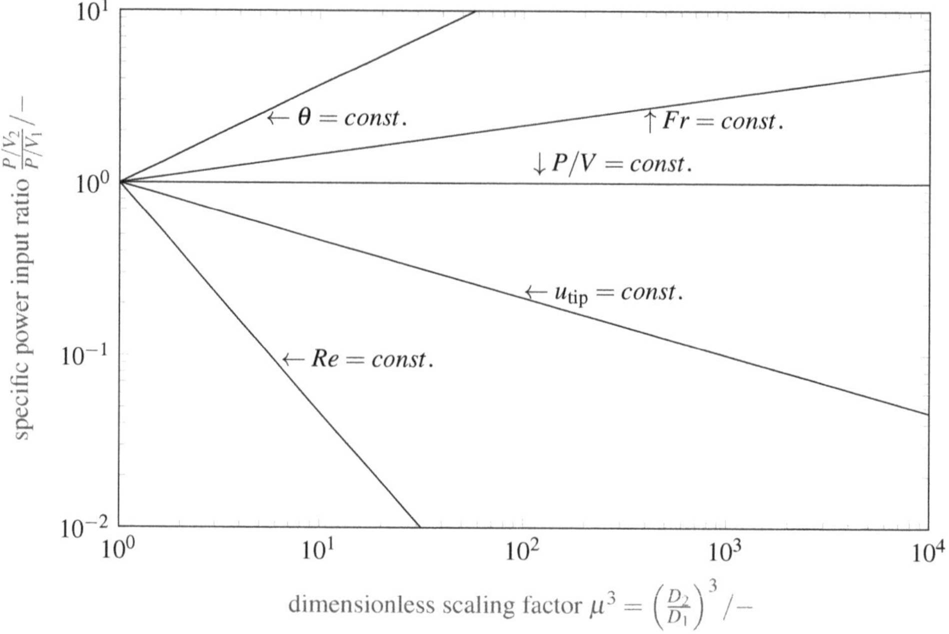

Figure 1.1. Dependence of the specific power input during scale-up of stirred tank reactors on the basis of various constant operating parameters for different tank diameters $D_1 < D_2$, according to [Zeh02]

input over various scales, other important system parameters cannot be kept constant. For example, for a constant dimensionless mixing time θ, the specific power input P/V must increase significantly. However, if the stirrer tip speed u_{tip} is to be kept constant, the power input P/V must be significantly reduced with increasing scale. This impressively illustrates the difficulties of the scale-up.

Another dilemma is that although a large number of fundamental papers on the investigation and description of the operating behavior of stirred tank reactors have been published in recent decades and have thus made a significant contribution to the general state of the art today [Zlo01, Kra12, GO04, Mon15, Gez00, Mav01], even minor deviations from the investigated model geometries mean that the results in the literature can no longer be applied directly to other problems. This is shown particularly impressively by the example of the global mixing time of an aerated stirred tank reactor on an industrial scale by Rosseburg et al. 2018 [Ros18], because on the one hand, there is a lack of publications and thus of correlations to the mixing time behavior of stirred tank reactors on industrial scale, and on the other hand the geometric installations often do not correspond to the own scale-up. This leads to increased uncertainty during scale-up, especially for multiphase systems, whose operating parameters have complex interactions.

Furthermore, the classical characterization approaches for describing aerated stirred tank reactors usually use integral measured variables, such as the global mixing time or the global volumetric mass transfer coefficient. This method of characterization corresponds to the Eulerian approach. The Eulerian description refers to a stationary coordinate system for the description of physical processes. Here the observer is at a fixed location for the description of the environment. Whereas in the Lagrangian approach, physical properties are described using a moving observer, for example by a particle in a flow [Her06].

Since the motivation of this work is the further development of established measurement methods as well as the better understanding of heterogeneities in stirred tank reactors, the following topics are discussed. First, the required theoretical fundamentals about power input and mixing are given for the general description of the operating behavior of stirred tank reactors. Following this, the Lagrangian analysis methods are described, which are used to describe the mixing structures as well as the heterogeneities in the system. Furthermore, a transfer from the Eulerian to the Lagrangian description is shown, which provides deep insights into the prevailing transport processes in stirred tank reactors.

Due to the fact that local measurement methods for the flow field measurements and mixing time measurements used in this thesis require a high experimental effort, as well as a perfect and non-distorted optical access, the methods in this thesis are implemented and tested on a stirred tank reactor at laboratory scale. First, classical power input and mixing time characteristics are measured on the laboratory-scale stirred tank reactor and compared with the literature. This measuring procedure is used to validate the experimental setup and the used measurement methods. Subsequently, spatially resolved mixing time experiments are performed, and the results are discussed. Finally, Lagrangian particle tracking velocimetry is used to determine the local and three-dimensional flow field, and Lagrangian analysis of the local residence time and dispersion are shown. Following the laboratory scale results, an outlook on the extension to industrial scale stirred tank reactors is given.

2. State of the Art and Basic Knowledge

In the following, the fundamental theories as well as the state of the art on the design and characterization of aerated stirred tank reactors are described. First, the general design of aerated stirred tank reactors is discussed. Then, the fundamentals of power input and mixing time are discussed in more detail. Finally, Lagrangian approaches to describe the single-phase transport processes are described and discussed.

2.1. Reactor Design and Characterization

The following subsections first describe the fundamentals of stirred tank reactor design and the most important characterization parameters. On the one hand, typical geometric designs and their desired influence on the operating behavior of stirred tank reactors are discussed, and on the other hand, the problem-specific nomenclature is defined so that no misunderstandings occur. Subsequently, the theoretical fundamentals of specific power input, mixing time and Lagrangian analysis methods with respect to stirred tank reactors are explained.

2.1.1. Fundamentals of Stirred Tank Reactor Design

In principle, there are virtually no limits to the design of stirred tank reactors for mixing and homogenization tasks. Essentially, a stirred tank reactor consists of the basic elements of the stirred tank reactor, the stirrer shaft including stirring elements, a motor, baffles and a temperature unit. Moreover, the design of stirred tank reactors is standardized in DIN 28130 [Nor07]. However, the greatest design diversity depends on the dimensions of the individual components and their respective placement [Kre02]. Furthermore, a gas supply unit is required for multiphase gas/liquid systems. Based on the listed basic components, it becomes clear that a large number of different designs are possible due to the free combination and placement of the components in the system. Especially due to this variety of design possibilities, it is very difficult to find a comparable configuration in the literature that is suitable for new problems. For this reason, the most important theoretical fundamentals are explained and described in the following, using one of the configurations used in this work as an example for each case.

Figure 2.1 i) shows the basic components which are usually used for the design of stirred tank reactors. However, the number and the shape of the baffles and stirrers in the system can be freely selected. Baffles are used to intensify the turbulence and thus the power input in the system in the case of low viscosity media, as well as to avoid the formation of a trombe. In general, 3-4 baffles are used. In the case of 4 baffles, the system is also referred to as a fully baffled system [Zlo01]. The main dimension quantities are shown in Figure 2.1 ii). Basically, a fill height to reactor diameter ratio $H/D_{\mathrm{STR}} = 1$ is desined for standard processes.

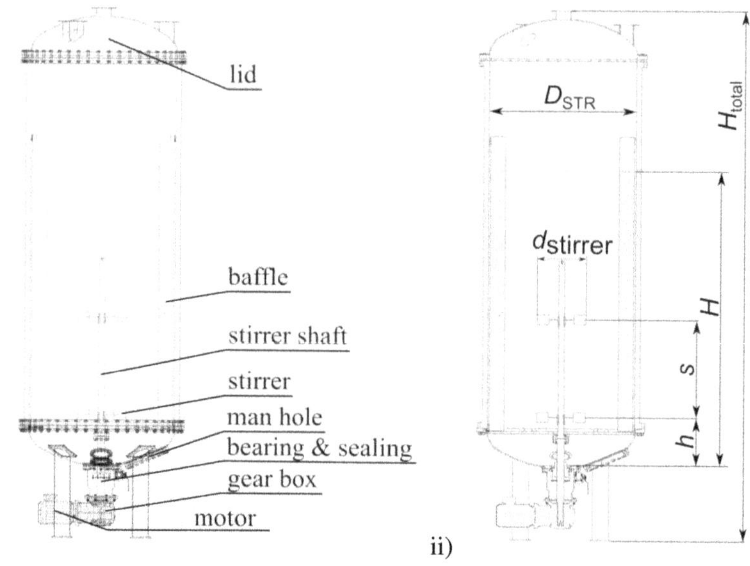

Figure 2.1. i) Sketch of the essential components of a stirred tank reactor; ii) main dimensions in relation to stirred tank reactors and their components

However, the tank diameter quickly becomes larger $D_{\mathrm{STR}} > 4.6\,\mathrm{m}$ with increasing reactor volume [Zlo01], which limits tank size due to transport constraints. Due to this, for industrial scale stirred tank reactors with a volume $V_{\mathrm{STR}} > 2\,\mathrm{m}^3$, a filling height to reactor diameter ratio $H/D_{\mathrm{STR}} > 1$ is used. However, this also results in disadvantages which have to be solved by changing the design. On the one hand, it is recommended to distribute several stirrers over the reactor height H for slim vessels. This means that the stirrer shaft must be longer and thus more stable. On the other hand, the mixing time increases accordingly with increasing filling heights to reactor diameter ratio $H/D_{STR} > 1$ [Zlo01]. This must either be taken into account in the design or compensated for by alternative stirrer combinations.

For different purposes, a variety of stirrer types can be selected. In principle, the different types of stirrers can be divided into two categories. A distinction is made between axially and radially pumping stirrers. Figure 2.2 shows a typical radially pumping Rushton turbine (i)) and an axial pumping pitched blade stirrer (ii)).

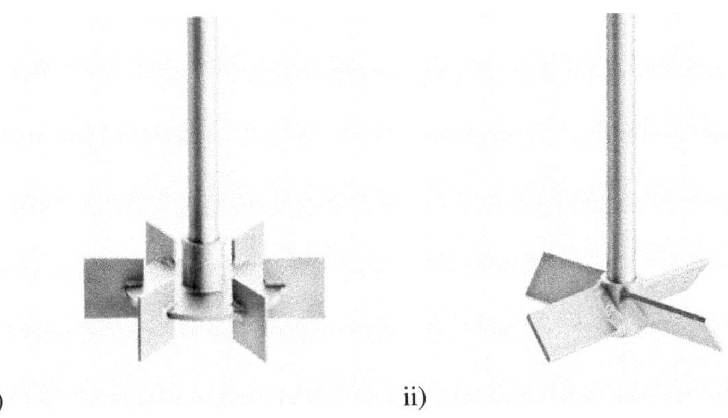

Figure 2.2. Representation of the characteristic axially and radially pumping stirrer: i) Rushton turbine, ii) pitched blade turbine; photos from [Gmb21]

The predicted flow structures resulting from the use of the two stirrer types are shown in an idealized form in Figure 2.3. The structures shown refer in each case to an axially and a radially pumping stirrer type, as well as the combination of two of the respective stirrer types. The vortices shown as examples in Figure 2.3 start at the lower stirrer. The direction of rotation of the vortices is indicated by an arrowhead at the end of the vortices. Figure 2.3 i-ii) shows the predicted large-scale flow field of one or the combination of two axially pumping stirrers. The figure clearly shows that the liquid is drawn to the top of the stirrer and flows off axially. Subsequently, the liquid is redirected in the reactor bottom and deflected upwards in the area near the wall of the cylindrical reactor. Furthermore, it is obvious that the influence of the stirrer on the upper part in the system is not particularly large. In order to achieve a more uniform influence by the stirrers over the entire reactor volume, several stirrers are often placed on one stirrer shaft (see Figure 2.3 ii)).

The behavior of radially pumping stirrers like a Rushton turbine can be seen in Figure 2.3 iii). In radially pumping stirrers, analogous to centrifugal pumps, the liquid is sucked in the area of the axis of rotation and then transported radially outward. The volume moved

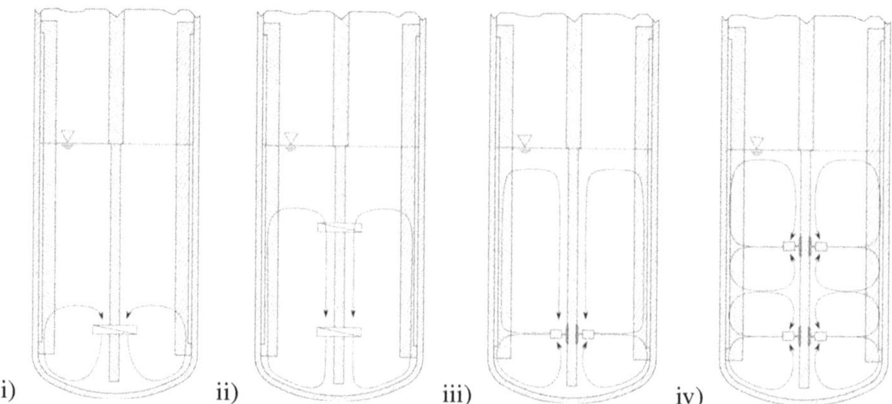

Figure 2.3. Representation of the characteristic flow behavior of a: i) single pitched blade turbine (down pumping), ii) double pitched blade turbine (down pumping), iii) single Rushton turbine and iv) double Rushton turbine

radially by the stirrer is then discharged downward or upward at the inner reactor wall. As a result, two characteristic vortices are typically formed in radially pumping stirrers. Due to this, the volume that is primarily influenced by the stirrer is larger than in comparable axially pumping stirrers. Nevertheless, even with radially pumping stirrers, several stirrers are often installed on the shaft to homogeneously distribute the power introduced into the system over the reactor height (see Figure 2.3 iv)).

In addition to the subdivision according to the main pumping direction, other important properties can be assigned to stirring elements. Table 2.1 provides an overview of which agitators are suitable for which process engineering tasks. Deciding which type of stirrer should be used is largely determined by the process engineering task. At this point, a distinction is made between suspending and dispersing two or more phases. The tasks are listed line by line in the first column. In the other columns, the respective basic stirrer types are listed and their suitability for the corresponding tasks is indicated. In addition to the process engineering task, the expected viscosity of the medium to be stirred is also important for the selection of the stirring elements because the range of influence of the stirrer decreases with increasing viscosity, so that the stirrers are consequently designed to be larger (see table 2.1 anchor and helical stirrer).

Furthermore, not all stirrer types are suitable for dispersing a gas/liquid or solid phase in an aqueous phase. For the dispersion of a second phase, high shear fields are required. Radially pumping stirrers are particularly suitable for this purpose. A special case is suspension,

Table 2.1. Selection guide for suitable basic stirrer types for various mixing and dispersing tasks, based on [Kre02]

typical stirrer	propeller	Rushton turbine	dissolver	pitched blade	anchor	helical
example						
pumping mode	axial	radial	radial	axial radial	radial	axial
homogenize	+	-	-	+	+	+
disperse gas/liquid	-	+	-	-	-	-
disperse liquid/liquid	-	+	+	-	-	-
disperse solid/liquid	-	-	+	-	-	-
heat transfer	-	-	-	+	+	+
viscosity range [Pa·s]	< 2			< 10		< 1,000
Preferred flow condition	turbulent			transition		laminar

which is a special form of solid/liquid dispersion, but the density of the solid is significantly greater than that of the liquid and thus the solid tends to sediment [Kre02].

As already mentioned, baffles prevent the formation of a trombe and increase the degree of turbulence in the system. They contribute to the intensification of the stirring task and thus to mass transfer in the system. Regarding heat transfer, for small amounts of heat to be removed or added, a simple double jacket reactor can be used, where the cool/hot medium circulates in the double jacket and controls the temperature in the stirred tank reactor via an external heat exchanger. For higher heat quantities to be removed or supplied, additional heating/cooling elements can be installed in the stirred tank reactor to intensify the heat exchange. Especially for temperature-sensitive processes, such as in the biotechnology industry, internal heat exchangers can improve the heat control by providing a significantly larger surface area compared to the double jacket [Zlo01]. However, the large surfaces of internal heat exchanger bundles also pose an increased risk of contamination, hence internal heat exchangers are not used in critical processes.

2.1.2. Fundamental Theory of Power Input

The following details on the power characteristics of stirred tank reactors refer only to single liquid phase operating. The theoretical fundamentals of the global power input in stirred tank reactors are discussed, followed by the fundamentals of local description of energy and turbulence by means of energy spectra.

Global Power Input

The following explanations are based on a dimensional analysis according to the Π-theorem [Zlo01]. With respect to the power input to be characterized, the power input P is set as the target value in the dimensional analysis. Figure 2.4 shows those parameters which have an influence on the power input. These parameters can be divided into geometrical (vessel diameter D_{STR} and stirrer diameter d_{stirrer}), physical (viscosity η and density ρ) and process-specific properties (stirrer frequency n). The Π-theorem states that the correlation of

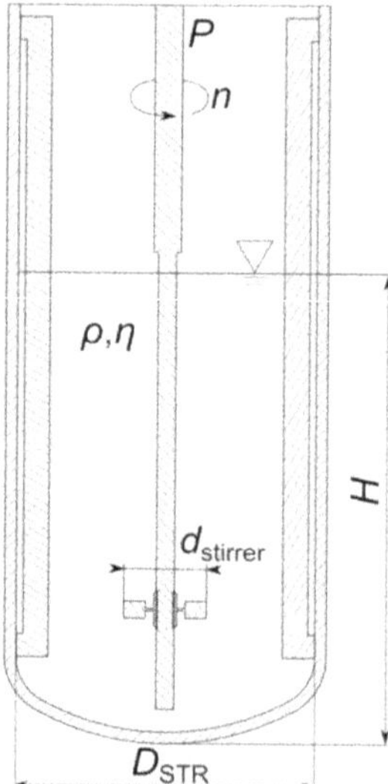

Figure 2.4. Main dimensions of stirred tank reactors and main unaerated properties; tank diameter D_{STR}, stirrer diameter d_{stirrer}, viscosity η, density ρ, stirrer frequency n and power input P

α physical quantities characterizing a system can be reduced by γ dimensionless numbers . The number

$$\gamma = \alpha - \beta \tag{2.1}$$

of dimensionless numbers results from the number of basic dimensions β and the number of physical quantities α. Based on Figure 2.4, five independent physical quantities with three individual basic dimensions can be identified, which have a direct influence on the power input. Therefore, the power input can be described by two different dimensionless numbers.

One of the essential and most important parameters for describing stirred tank reactors is the first dimensionless number

$$\Pi_1 = \frac{P_{\text{stirrer}}}{\rho \cdot n^3 \cdot d_{\text{stirrer}}^5} = \frac{2 \cdot \pi \cdot M_{\text{stirrer}}}{\rho \cdot n^2 \cdot d_{\text{stirrer}}^5} = Po \tag{2.2}$$

with the stirrer frequency n, the liquid density ρ and the stirrer diameter d_{stirrer}. The first dimensionless number Π_1 is also called the power number Po. In addition to the power number, the second dimensionless number Π_2, the stirrer Reynolds number

$$\Pi_2 = \frac{n \cdot d^2 \cdot \rho}{\eta} = Re_{\text{stirrer}} \tag{2.3}$$

with the stirrer frequency n, the stirrer diameter d and the dynamic viscosity η is a very important quantity for describing the fluid dynamics in stirred tank reactors.

In this case, the power input into the system is balanced via the torque present at the stirrer shaft. As a rule, the stirrer power cannot be determined directly at the stirrer shaft. Due to design limitations, the stirrer power

$$P_{\text{stirrer}} = P_{\text{total}} - P_{\text{bearing}} = 2 \cdot \pi \cdot n(M_{\text{total}} - M_{\text{bearing}}) \tag{2.4}$$

is usually determined between the motor and the stirrer shaft bearing with the stirrer frequency n and the respective torque M. If the torque on the shaft is measured between the motor and the shaft bearing, the bearing torque must be subtracted from the measured torque. In this case, if possible, the idle torque of the bearing must be known or measured.

Furthermore, the power input depends on various geometrical factors, such as the stirrer diameter d, the installation height h, the stirrer type, the number of stirrers N and the reactor geometry (diameter D, bottom shape). A typical visualization of the stirrer performance characteristics is given in Figure 2.5. In Figure 2.5 the power-number for three stirrer types is plotted against the stirrer Reynolds number. The visualization in Figure 2.5 clearly shows that the power characteristics can be divided into three main regimes. For very small Reynolds numbers ($Re < 10$), the power-number in the laminar flow region decreases reciprocally

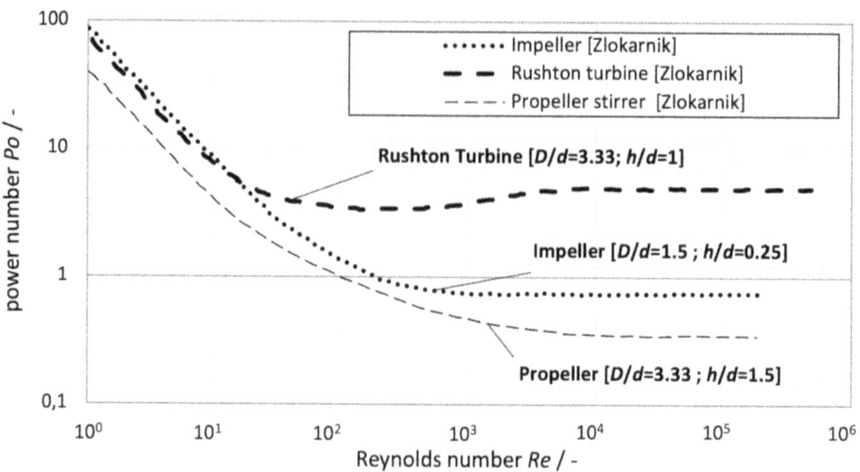

Figure 2.5. Visualization of the dependence of the power-number *Po* on the stirrer Reynolds number Re_{stirrer} for three different stirrer types; visualization based on [Zlo01]

with increasing Reynolds numbers. From a typical Reynolds number of $Re \gg 10,000$), the power-number remains constant, so it can be assumed that a fully turbulent flow prevails in the stirred tank. In addition, each stirrer or each individual stirrer/reactor configuration has its own stirrer performance characteristics, so that those stirrer performance characteristics should be known in particular for the transfer or scale-up of stirred tank reactors. Hudcova et al. and Fitschen et al. have shown that the combination of several stirrers does not necessarily lead to a direct doubling of the power input, since this is directly dependent on the distance between the stirrers [Hud89, Fit21]. For a stirrer spacing of $s > 2$, it can be assumed that the cumulative power input is twice that of the single stirrer, or the sum of both power numbers *Po* for the combination of different stirrers.

Due to the fact that the power input into the system is inferred from the stirrer power, the power number is an integral quantity that describes the average power input for the entire system. For this reason, the specific power P/V is often given to describe process engineering processes, in which the input power is related to the stirred volume. Another well-known characterization parameter is the average energy dissipation rate

$$\tilde{\varepsilon} = \frac{P}{V} \cdot \rho^{-1}, \tag{2.5}$$

which relates the power input to the mass rather than to the volume.

The Energy Transport in the Frequency Domain

The description of the system specific stirrer power characteristics by means of the power number is an important and frequently used method of characterization. However, no detailed or locally resolved insight into the energy introduced into the system can be determined on the basis of the power-number *Po*. Especially for shear sensitive processes, like the cultivation of CHO cells in the bio-pharmaceutical industry, locally resolved energy dissipation rates can help to better design the processes and the apparatus. Therefore, maximum values should be avoided in terms of frequency and strength. For this purpose flow field measurements with high spatial and temporal resolution can be performed by particle image velocimetry, and the absolute and/or turbulent kinetic energy can be calculated on the basis of such Eulerian data.

The following theoretical considerations are based on statistical considerations for the description of turbulence by G. I. Taylor and H. Robert [Tay35, Kra64]. The most important consideration is that a homogeneous isotropic turbulence is assumed. This simplification is important, because for a homogeneous isotropic turbulence, the local spatial velocity fluctuation $\bar{u'^2} = \bar{v'^2} = \bar{w'^2}$ is the same for all directions. Thus, there is no preferred direction, and therefore, there is no dependence on the observation location in the description of the turbulence. However, turbulent structures are not detectable with this approach. The description of the turbulent structures can be performed in the time as well as in frequency domains. Either the lifetime of the structures is needed as a characteristic length scale for the analysis in the time domain, or the wavenumber k is needed for the analysis in the frequency domain. The wavenumber

$$k = 2 \cdot \pi / L \tag{2.6}$$

is defined by the length L of correlating structures and the conversion factor 2π. Figure 2.6 shows an example of the idealized vortex decay. The so-called vortex hypothesis of Lewis F. Richardson [Ric22] states that the largest vortices in the system are unstable and break up, forming several smaller vortices, thereby transferring the kinetic energy of the flow to smaller scales. The larger structures are defined by the geometrical boundary conditions of the considered system and describe the scale at which the energy is added to the system. In the example of a stirred tank reactor, this would be the stirrer diameter. Above a certain maximum wave number, the vortex structures cannot decay any further and the kinetic energy conserved in the structures dissipates via viscous friction to thermal energy.

The total kinetic energy

$$E(t) = 0.5\left(u(t)^2 + v(t)^2 + w(t)^2\right) \tag{2.7}$$

can be calculated based on the three-dimensional velocity fields vector $\boldsymbol{v}$ with the respective velocity components $u(t)$,$v(t)$ and $w(t)$ [Gab09]. For the calculation of the energy cascade, the total kinetic energy

$$\hat{E}(\boldsymbol{k},t) = 0.5\left(u(\boldsymbol{k})^2 + v(\boldsymbol{k})^2 + w(\boldsymbol{k})^2\right) \tag{2.8}$$

in the frequency domain is calculated in the same way as in the time domain [Ale18]. Based

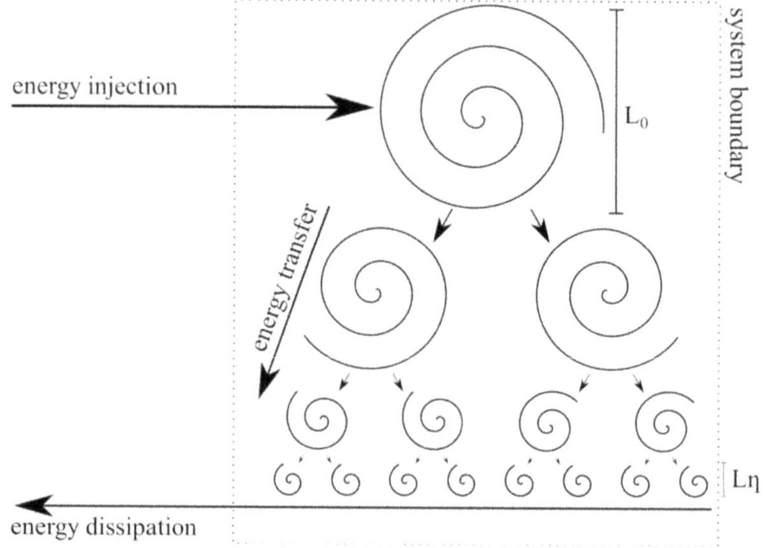

Figure 2.6. Example representation of the energy cascade using the example of vortex decay and the energy transfer over the scales

on a spectral analysis of the total kinetic energy in the frequency domain, the distribution of kinetic energy in a turbulent system can be calculated. Here, the energy spectrum $\hat{E}(k)$ in the frequency domain describes the total kinetic energy

$$E(k,t) = \iiint \hat{E}(\boldsymbol{k},t)\mathrm{d}\boldsymbol{k} \tag{2.9}$$

for different scales $1/k$ [TKN19] over the respective spherical shell of width $\Delta k = 2/L$. The scales are represented by the characteristic lengths L of the vortices in the system. The sum of all spherical shells obtains the total kinetic energy

$$\varepsilon(t) = \Delta k \sum_k E(k,t). \tag{2.10}$$

Assuming a three-dimensional isotropic turbulence, the energy spectrum according to KOLMOGOROV's theory is shown in Figure 2.7. Figure 2.7 shows an example of the energy

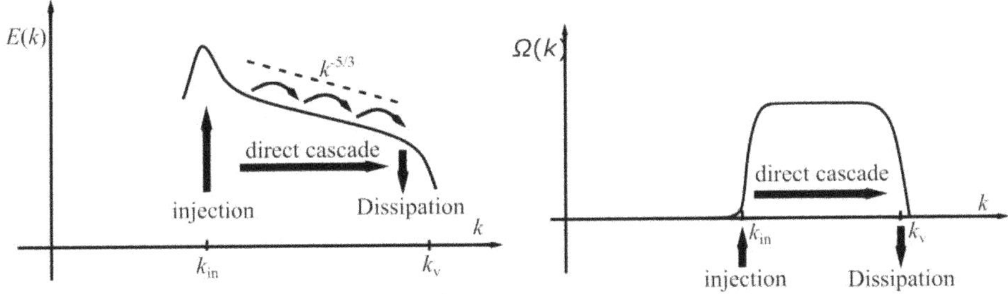

Figure 2.7. Representation of the energy cascade (left) and the energy flux (right) of the absolute kinetic energy in the frequency domain [Ale18]

spectrum $E(k)$ (left) and the energy flux $\Omega(k)$ (right) in double logarithmic scale. The energy spectrum can be assigned to three different spectral ranges:

energy containing: $k \sim k_{\text{in}}$

inertial range: $\hat{E}(k) \sim k^{-5/3}$; at $k_{\text{in}} \gg k \ll k_\upsilon$

dissipation range: $k \sim k_\upsilon$

Around the range of larger scales (small wavenumber) the kinetic energy at $1/L_0$ is added to the system and conserved. With increasing wavenumbers the vortices become smaller and the total kinetic energy of each vortex size decreases reciprocally by -5/3 according to KOLMOGOROV's theory. Above a characteristic wavenumber, the KOLMOGOROV's length scale η, there is no more energy transfer and the kinematic energy dissipates only via viscous friction. Moreover, the function of the spectral distribution of the absolute kinetic energy

$$\hat{E}(k) = C_\text{k} k^{-5/3} \varepsilon^{2/3} \tag{2.11}$$

can be described by the universal Kolmogorov constant $C_\mathrm{K} \approx 1.5$, the wavenumber k and the energy dissipation rate ε. Equation 2.11 is based on a dimensional analysis by Kolmogorov for isotropic turbulence and has been proven several times [Bak08].

The energy spectrum $\hat{E}(k,t)$ can be used to describe important properties of the system. On the one hand, the structures can be identified with which one has the highest kinetic energy. On the other hand, the energy flux and thus the scale transfer between the structures can be described. On the basis of these data, the local kinetic energy as well as the energy dissipation rate $\hat{\varepsilon}(k)$ can be described for those scales which are, for example, in the order of magnitude of animal cells, depending on the operating parameters and the design of different reactor geometries.

2.1.3. Fundamental Theory of Mixing Behavior in Stirred Tanks

The mixing time characteristic of a stirred tank reactor is an important design and operating parameter, especially for processes where two or more phases have to be continuously brought into contact over a long period of time (several days). For example, in the cultivation of microbial or animal cell processes, nutrients must be added continuously in addition to the supply of oxygen through active aeration of the reactor volume. Efficient mixing ensures that there are sufficients nutrient in the entire system over the entire process time. Furthermore, efficient mixing minimizes local concentration gradients (e.g. at high consumption rates). In the following, the basic principles of mixing time in stirred tank reactors are described first as a global and then as a local parameter.

Global Mixing

In principle, mixing in stirred tank reactors means the homogenization of two different phases (gas/liquid or solid/liquid) or even more than three phases, with the aim of equalizing concentrations in the system. For this reason, the mixing time is often defined by the degree of homogeneity in a system. In addition, the distinction between macro, meso and micro mixing must be taken into account. In stirred tank reactors, macro mixing depends mainly on the large-scale flow structures, such as large-scale eddy structures, dead zones or compartments [Man94, Sie11, Woz11]. Meso mixing depends on local flow structures, such as small-scale vortex structures or in the wake of gas bubbles. Thus, macro-mixing and meso mixing depend on the geometrical characteristics and the operating parameters of the stirred tank reactors. Micro mixing depends on the physical properties of the system and, therefore, on molecular diffusion.

A commonly used measure to characterize mixing performance is the mixing time. To determine the mixing time, usually the 95-% criterion is used [Kra02, Hib79]. Hereby, a system is considered mixed when 95-% of the target concentration is reached in the entire system. The determination of the time required for mixing is generally based on the measurement of an impulse response. For example, a local peak in conductivity is used to measure the conductivity over time in the system after the addition of a salt solution with a correspondingly higher conductivity.

In general, the measuring methods for determining the mixing time can be divided into chemical and physical methods. Physical methods use a local impulse to change the physical properties, such as temperature, conductivity or pH. The change in physical properties can be measured and recorded either via a local probe measurement technique or via imaging methods such as LIF (Laser-Induced Fluorescence) or tomographic methods [Mew02]. Local measurement methods are simple to implement, robust and easy to evaluate. However, local measurement methods have the disadvantage that the entire system is determined from the measurement at one point. This means that possible dead zones in the system cannot be detected. Imaging methods have the advantage that they are able to resolve the entire system, so that poorly mixed zones can also be detected. However, imaging methods are significantly more extensive in handling and evaluation than measurements with local probes.

The chemical methods are also often called optical methods because the color of the medium in the stirred tank reactor changes based on a chemical reaction. The chemical reaction is usually evaluated optically. But, this also requires optical access to the system, which cannot always be guaranteed. With the chemical method, a distinction is made between single and multiple transition systems. The single transition systems include the redox reaction of sodium tiosulfate with iodine or the neutralization of sodium hydroxide solution with hydrochloric acid [Sie11]. A starch solution or the pH indicator phenolphthalein is used as an indicator. Systems using a simple color change are also referred to as the decolorization method. The advantage of the decolorization method is that possible dead zones are clearly visible, but an excess of iodine or hydrochloric acid cannot be detected [Fit21]. In order to obtain a broader spectrum of "decolorization", for example in the neutralization reaction, pH indicators such as bromothymol blue can be used instead of phenolphthalein. However, this increases the computational effort required for the evaluation.

Single Phase Mixing

For single-phase systems, a dimensional analysis yields the following relationship for the mixing time

$$\Theta = \frac{f(Re)}{n} \tag{2.12}$$

with the stirrer Reynolds number Re and the stirrer frequency n [Zlo01]. In single-phase systems there is a preferential dependence of the mixing time on the power input and the geometrical and physical properties [Zlo01]. Furthermore, the function $f(Re)$ describes the dependence of the mixing time on the Reynolds number and has a characteristic curve, which is shown as an example in Figure 2.8. The global dimensionless mixing time $\theta = n \cdot \Theta$

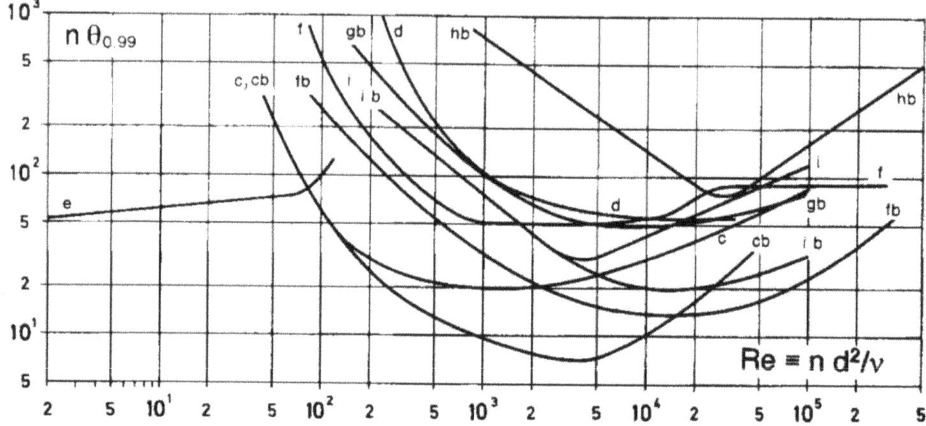

Figure 2.8. Dimensionless mixing time characteristic of different stirrers: a) cross beam, b) frame, c) blade, d) anchor, e) helical ribbon, f) MIG, g) turbine, h) propeller and i) impeller; +b = baffled tank [Zlo01]

is plotted against the Reynolds number Re for different stirrer types in Figure 2.8. In the range of small Reynolds numbers, the mixing time decreases as the square of the stirrer frequency. Depending on the stirrer type, there is a small range around $10^3 < Re < 10^4$ where the dimensionless mixing time reaches a plateau which means that the mixing time decreases linearly with increasing stirrer frequency. For very large Reynolds numbers, the mixing time is consistent, and the dimensionless mixing time increases quadratically with increasing stirrer frequency [Zlo01]. Additionally, a higher aspect ratio H/D for the same stirrer type causes the function $f(Re)$ to be linearly shifted to higher dimensionless mixing times [Kra02].

Multiphase Gas/Liquid Mixing

The mixing characteristics of a multiphase system, especially with very large differences in the density of the two phases, as in gas/liquid systems, are much more complex. For a better description, additional important parameters will be used. For this, two common ways of describing the absolute gas volume flow rate exist [Nau15]. The superficial gas velocity

$$w_g^0 = \frac{q_g}{A_{cross}} \tag{2.13}$$

relates the gas flow rate q_g to the reactor cross sectional area A_{cross}. Additionally, the volumetric gassing rate

$$vvm = \frac{q_g}{V_{reactor}} \tag{2.14}$$

links the gas flow rate q_g to the reactor volume $V_{reactor}$. In addition, two gas/liquid flow states have to be distinguished in the multiphase operation of stirred tank reactors. For this, the dispersing capacity of the stirrer is described as a function of the specific power input P/V and the gas flow rate q. Figure 2.9 shows an example of the gas/liquid distribution in the system for the impeller flooding and loading state. The first state illustrates impeller flooding (see 2.9 i)): the specific power input P/V is too low for the gas volume flow q and an axial overflow of the stirrer occurs. With increasing specific power input P/V or decreasing gas volume flow q, the gas phase can be distributed over the reactor cross-section. This condition is called impeller loading (see 2.9 ii)). The third state also shows the state of impeller loading (see 2.9 ii)), but the specific power input $P/V$is correspondingly large, so that gas recirculation occurs in the system [War86].

In case of gas/liquid systems, the general principle that the mixing time becomes shorter with increasing power input or Reynolds number also applies to multiphase systems. However, the mixing time characteristics of a gas/liquid system are special. This is due to the fact that the bubble rise and the flows induced by it (buoyancy driven flows) [Ros18], depending on the gas volume flow rate and the distribution of the gas phase over the reactor cross-section, contribute significantly to the mixing.

On the one hand, this leads to a significantly shortened global mixing time in the case of a heterogeneous gas distribution over the reactor cross-section (strong heterogeneous bubble-induced buoyancy driven flow). On the other hand, the effect of the bubble-induced buoyancy driven flow is significantly reduced in the case of a homogeneous gas distribution over the reactor cross-section. In this case, the bubble swarm rises uniformly and consequently does

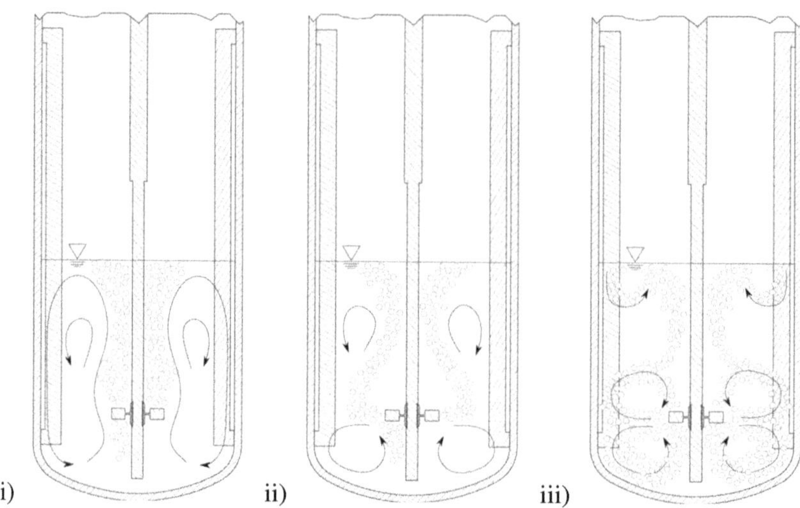

Figure 2.9. Visualisation of the characteristic impeller gas/liquid behavior of an aerated stirred tank with one single radial pumping stirrer: i) impeller flooding, ii) impeller loading, iii) impeller loading with gas recirculation, according to Warmoeskerken 1986 [War86]

not contribute to an improved mixing [Ros18].

In Figure 2.10 the global mixing time Θ is plotted against the Reynolds number Re for an industrial scale aerated stirred tank reactor [Ros18]. In the range of small Reynolds numbers and aeration, the global multiphase mixing time is much smaller than the single phase mixing time. In the Range of around $2 \cdot 10^5 < Re < 4 \cdot 10^5$ the multiphase mixing time increases with increasing Reynolds number depending on the aeration rate until the multiphase and the single phase mixing time values collapse at high Reynolds numbers. The influence of the gas phase on the global mixing time can be explained by bubble-induced buoyancy driven flows [Ros18], depending on the aeration rate. In case the stirrer power is not sufficient to distribute the gas phase homogeneously over the cross-section at a fixed aeration rate, this results in a very strong local upward flow due to the rising bubbles. These strong and local buoyancy driven flows dominate the large-scale flow structures in the system and are able to break up compartments in the system. As shown in Figure 2.3, compartments are formed either between two radial stirrers (see Figure 2.3 iv)) or by the fact that for a high aspect ratio of $H/D > 1$ only a part of the system is stirred (see Figure 2.3 i)).

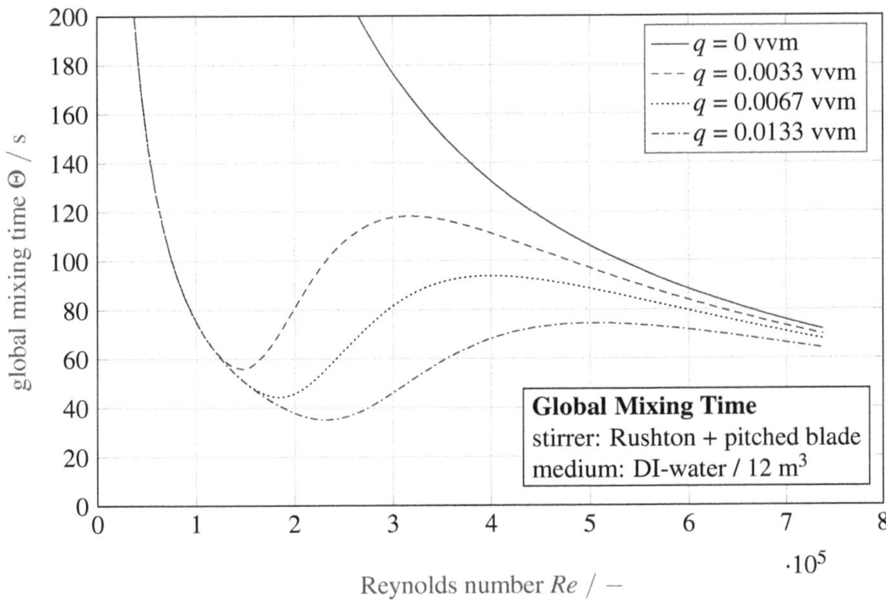

Figure 2.10. Global mixing time characteristic of a 12 m^3 aerated stirred tank reactor (q: gas flow rate)[Ros18]

Based on the data of [Ros18], it can be concluded that, especially in two-phase mixing, the flow regime (flooding /loading) and the distribution of the gas phase over the reactor cross-section have a major influence on the mixing time. This is of particular importance in fed-batch processes, since the flow regime can shift over time due to the changing liquid volume and aeration rate, and therefore a constant mixing time cannot necessarily be assumed.

For a better understanding of the conditions under which coherent compartments in the system dominate the mixing time or how the barriers between two compartments can be broken down, the description of local mixing is given below.

Consideration of Local Mixing Performance

Based on the differentiated consideration of macro, meso and micro mixing [Som10] and the generally accepted fact that for very large Reynolds numbers the mixing time is constant [Zlo01], the concept of local mixing is described and discussed in more detail below. Furthermore, the term 'local' or 'global' does not necessarily refer to the distinction between micro and macro mixing. The distinction between local and global mixing refers to the framework for which the mixing time is described. The global mixing time is a mean value for the entire system and, therefore, a mean of all local measured mixing times. Additionally, a subdivision into macro or micro mixing can be made for a local as well as global description of mixing. Which mixing scale will be considered in the following and will be explicitly stated in the results.

The knowledge that for very large Reynolds numbers the mixing time is constant [Zlo01] in stirred tank reactors can be explained by the fact that macro mixing cannot be further improved. Thus, the global mixing time depends only on the micro mixing, which by definition depends on the material properties and cannot be influenced by the operating parameters (except for a temperature increase) [Bou83].

Assuming that the physical properties of a system do not change during mixing, the characterization of the system based on the macro mixing time is more precise, because the selective measurement of the macro mixing of a system allows the geometric system to be characterized as a function of the operating parameters. However, a differentiated consideration of the mixing time only makes sense if it can be assumed that the process is not limited by mixing, as for example it is the case with very fast reactions.

Figure 2.11 shows the global mixing Θ time of a stirred tank reactor as a function of the mixing ratio X for an acid/base reaction, where a pH indicator is decolorized in the basic range by the addition of an acid. It can be shown that the global mixing time Θ decreases reciprocally with an increasing excess of acid and is nearly constant from a ratio of $X > 2:1$. The reciprocal behavior of the mixing time is attributed by the authors to the fact that for a 1:1 mixing ratio, complete decolorization only occurs when the concentrations in the entire system are balanced. The global mixing time for 1:1 mixing is thus composed of macro and micro mixing. Whereas with a mixing ratio greater than two, the macro mixing dominates, since here the concentrations do not have to be balanced in the entire system for complete decolorization to occur. Rather, with a mixing ratio greater than two, only the large-scale distribution of the added acid is shown.

For a better description of the subdivision into macro and micro mixing, Figure 2.12 shows the distribution of a tracer (yellow points) in a stirred tank reactor for 1:1 (first line) and 2:1 (second line) mixing. Based on the simplified representation in Figure 2.12 , it can be seen that in the case of 1:1 mixing, there must necessarily be a complete concentration

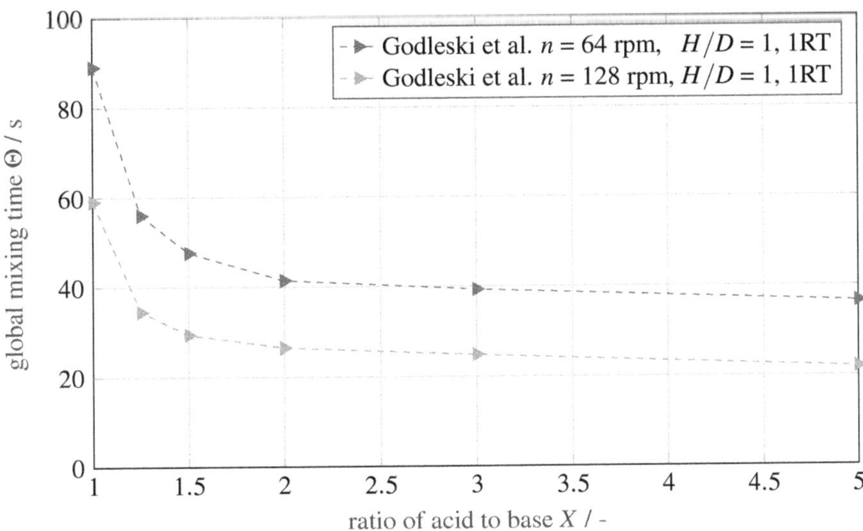

Figure 2.11. Comparison of the global mixing time with data for different mixing ratios of acid to base, according to Godleski et al. [God62]

balance in the system for complete decolorization to be evident. In the case of 2:1 mixing, only an excess must be present at each location in the system for the entire system to be decolorized. Nevertheless, in both cases the degree of homogeneity $M = 1$ and thus the equilibrium of concentrations in the whole system is achieved at the same time. However, due to the fact that in the case of 2:1 mixing the system is already decolorized, or the color change is completed, only the macro mixing can be measured. Thus, it should be noted that the degree of homogeneity in the system cannot be described with the macro mixing method. Figure 2.13 shows the chronology of the decolorization experiment using the pH indicator bromothymol blue. For automated evaluation, a grayscale image is calculated from the images and an objective representation of the mixing processes is obtained by evaluating the gray value progression during the experiment for each point in the image. For this purpose, the average gray value

$$I_{\text{avg}}(t) = \frac{1}{X \cdot Y} \sum_{x=1}^{X} \sum_{y=1}^{Y} (I(x,y,t) - I(x,y,t=0)) \tag{2.15}$$

is determined by means of averaging all pixels (*x*,*y*) of the reactor for each recorded time step, where x and y indicate the respective pixel position in the image and X and Y represent the maximum local resolution. The initial value $I(x,y,t=0)$ is subtracted from the gray

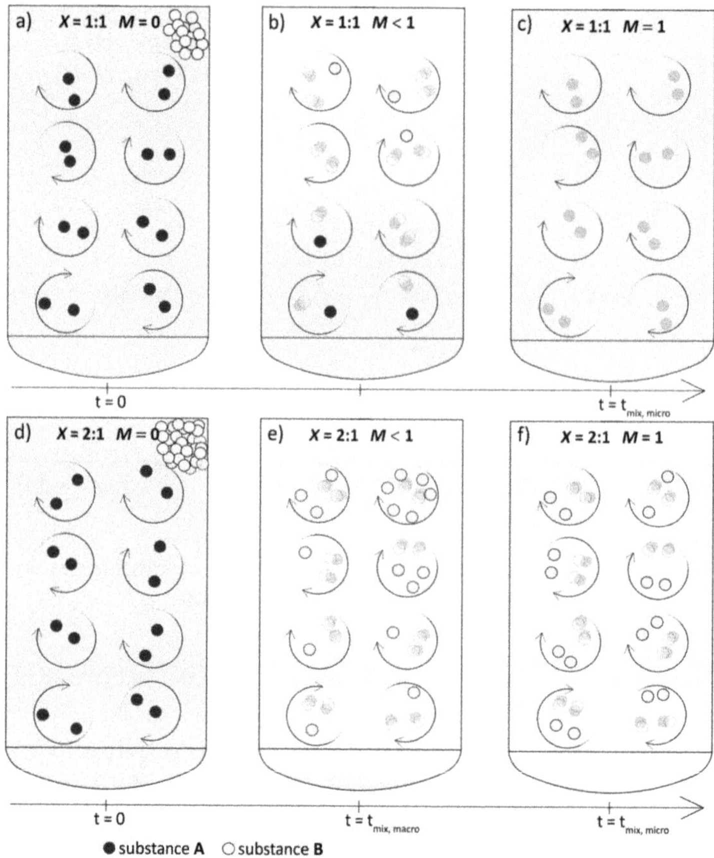

Figure 2.12. Micro vs. macro mixing - Schematic representation of macro and micro mixing of two substances A and B with the same substance mixture (upper row) and a double excess of substance B (lower row); M: degree of homogeneity [Fit21]

value $I(x,y,t)$ to exclude those pixels that are not part of the reactor. The gray value profile $I_{\text{avg}}(t)$ will then be normalized to the final value so that the mixing process runs from 0 to 1

$$I_{\text{avg,norm}}(t) = \frac{I_{\text{avg}}(t)}{I_{\text{avg}}(t=\infty) - I_{\text{avg}}(t=0)} \tag{2.16}$$

The mixing time criterion is set at 5 % deviation of the final value, as it is often described in the literature [Ros18]

$$\Theta_{0.05} = t(I_{\text{avg,norm}}(t) = 1 \pm 0.05). \tag{2.17}$$

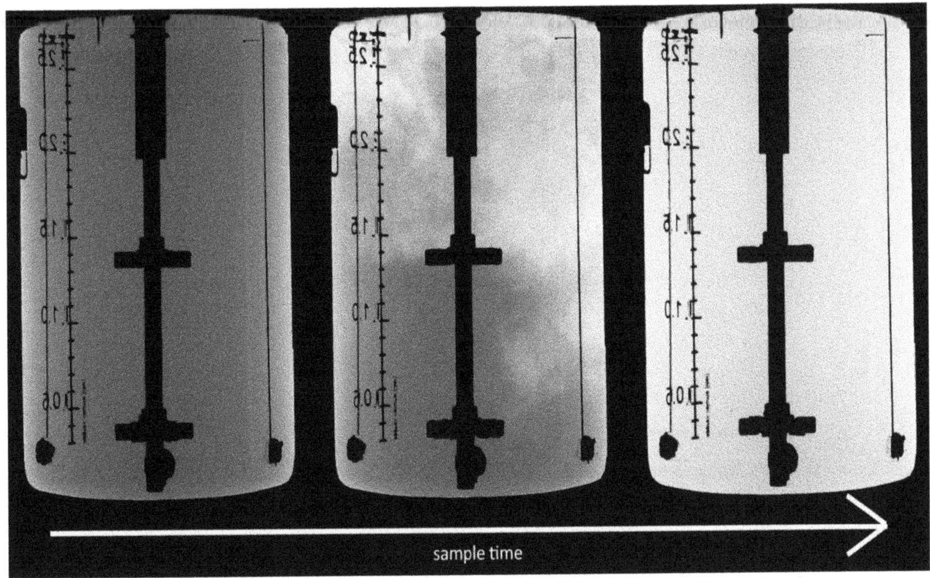

Figure 2.13. Decolorization process with bromothymol blue at time $t = 0; 15; 30$ / s (masking around the stirring elements and baffle holder)

With this method only the global mixing time can be evaluated objectively. The visual representation of the mixing process as well as the flow pattern remains only a subjective evaluation as long as the evaluation is not computer-aided. Often, this is represented by individual frames at certain points in time or as a time series of individual frames from the mixing process like in Figure 2.13 or described in [Alv02, Cab07, Sie11].

Instead of averaging the gray value information over all pixels in the system to determine the global mixing time, the gray value progression for each pixel can be evaluated with respect to the mixing time as in [Rod13, Fit21]. This allows the calculation of the theoretical relationship between the acid/base ratio shown in Figure 2.12 to be applied specifically to the measurement method in order to visualize the history of the mixing. Furthermore, this enables the global mixing time and the local mixing time to be determined simultaneously. However, it requires a much higher computational power (depending on the spatial and temporal resolution). Figure 2.14 shows an example of the normalized gray value curve for two different locations in a closed system, such as in a stirred tank reactor. Based on mixing phenomena on different scales, it is now possible not only to describe local mixing times and thus, for example, to identify poorly mixed zones, but also to describe the local gradients of the mixing. The higher the local gradients, the worse the exchange between two areas. To

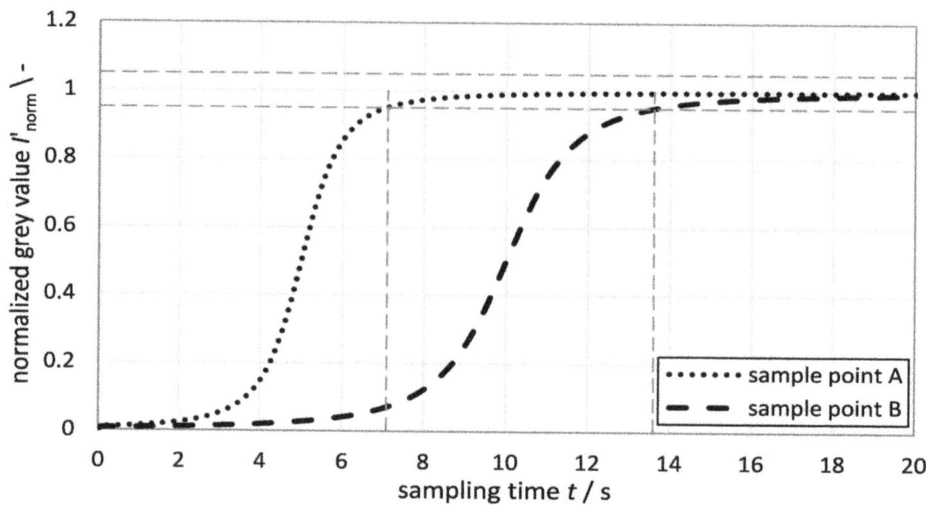

Figure 2.14. Profile of a cumulated residence time distribution of a step response at two different measuring points

determine the local mixing time, the normalized gray value

$$I'_{\text{norm}}(x,y,t) = \frac{I(x,y,t)}{I(x,y,t=\infty) - I(x,y,t=0)} \tag{2.18}$$

for each pixel (*x*, *y*) is calculated for each point in time *t* [Fit21]. In addition, the temporal gradients for each pixel have to be preprocessed. For this purpose, a Gaussian filter is used to calculate the smooth gray value. This procedure greatly reduces the measurement noise that might further disturb the evaluation, especially for aerated conditions. Hence, the mixing time according to

$$\Theta_{0.05}(x,y) = t(I'_{\text{norm}}(x,y,t) = 1 \pm 0.05) \tag{2.19}$$

can be calculated analogously for each pixel [Fit21]. In order to achieve a better comparison of different operating conditions with different global mixing times, the relative local mixing time distribution

$$\Theta_{0.05,\text{rel.}}(x,y) = \frac{\Theta_{0.05}(x,y)}{\Theta_{0.05}} \tag{2.20}$$

can be calculated by relating the local mixing time distribution to the respective global mixing time $\Theta_{0.05}$. The formula states that the educts at a pixel are mixed if the gray value reaches a deviation of 5 % from the final value. Eventually, the calculated local mixing times can be displayed as pseudo-color images [Fit21].

In addition to a graphical visualization of the local mixing time, the frequency density distributions for the respective local mixing times can also be considered. The frequency distribution is defined as a listing of the frequencies of all classes of the observed values of a variable [Fre10]. The frequency distribution of a variable can be represented both absolutely and relatively. In the case of the local mixing time, bins of equal size are calculated for an interval width of $\pm 60\%$ of the mean mixing time (1% point), and each pixel in the image, for which a certain local mixing time can be assigned, is assigned to the corresponding class.

As an example, Figure 2.15 and 2.17 visualize the relative frequency density distribution for a monomodal as well as a bimodal distribution with different variances σ (gaussian distributions). A monomodal distribution of the relative local mixing time in the system means that an average relative mixing time is characteristic for the entire system. In this case, a narrower normal distribution indicates a more homogeneous local mixing time distribution in the system. Thus, the whole system can be considered as one compartment.

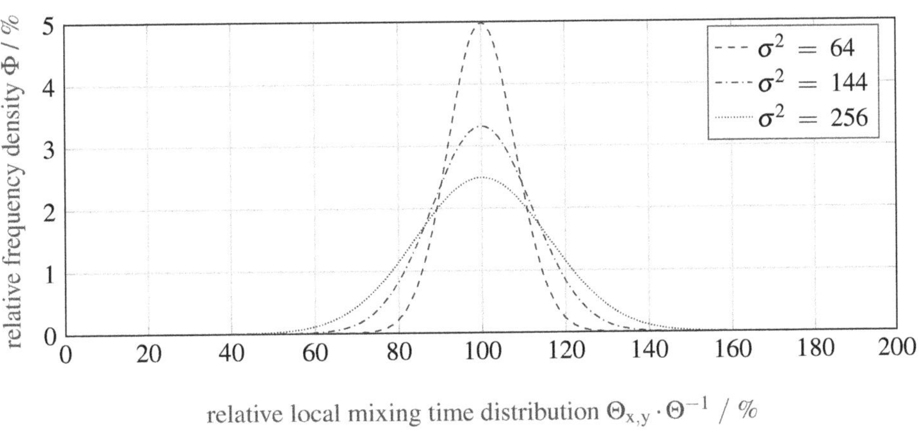

Figure 2.15. Representation of a monomodal probability density distribution with different distribution widths

A bimodal or multimodal distribution of the relative local mixing time in the system is an indication of the presence of two or more dominated compartments in the system. Compartments are coherent structures, which are characterized by a poor exchange with each other. Figure 2.16 shows three compartments for unaerated operating conditions and two radial stirrers. The compartment boundaries drawn are fictitious and are for illustration

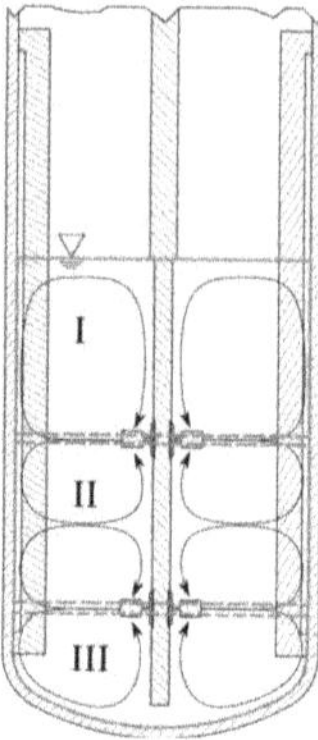

Figure 2.16. Exemplary representation of dominated compartments in a stirred tank reactor equipped with two radial pumping stirrers

purposes only. Figure 2.17 shows the relative frequency density distribution in the case of two dominated coherent compartments. The width of the respective maxima in the distribution, as

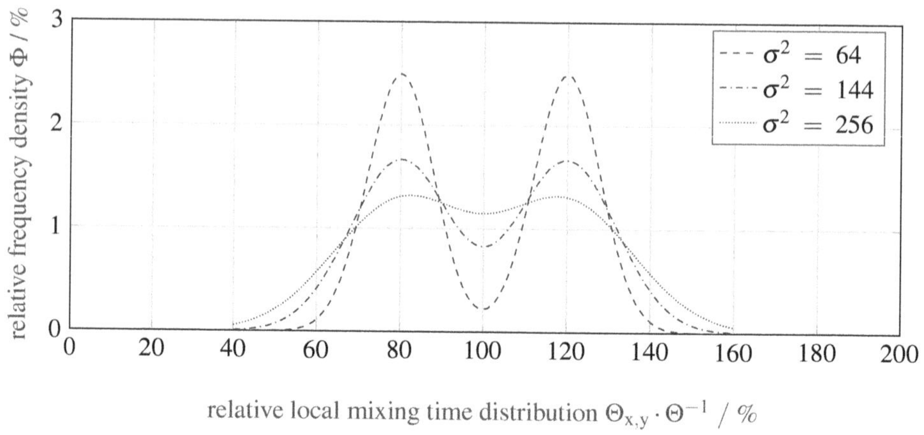

Figure 2.17. Representation of a bimodal probability density distribution with different distribution widths

well as the distance of the maxima from each other, are indications of the size and dominance of the respective compartments. Therefore, two compartments with low interaction would lead to a low variance but a large distance of the respective maxima from each other.

In summary, the consideration of the local mixing time distribution of a system is a useful tool, which provides significantly deeper insights into the transport processes. Not only the system as a whole can be described (global mixing time), but the history of the mixing (local mixing time) can also be visualized. The visualization of the local mixing time distribution is an Eulerian approach, which can be used not only to describe the local mixing, but also to classify the system into one or more compartments.

2.2. Fundamentals of the Lagrangian Evaluation Approach

In addition to the Eulerian approach for the description of problems of any kind, in which the observer has a stationary perspective, there is the Lagrangian approach. With respect to fluid mechanics problems, the observer takes the perspective of an infinitesimally small fluid element. However, with a few exceptions [Cam03], the Eulerian approach is preferably used to describe fluid mechanics problems. This is often due to the measurement techniques used, in which the measured variable is mapped onto an Eulerian grid (e.g. particle image velocimetry to determine velocity, energy dissipation or pressure fields or simple invasive probe measurements to determine concentration, temperature or pressure values in the system).

The classical and established measurement techniques for the description of process equipment are often based on the Eularian approach. However, Lagrangian analysis are already applied in a large number of studies describing transport and mixing phenomena in unsteady velocity fields. Especially in the field of oceanography and atmospheric research, many papers on the Lagrangian description of large-scale transport processes have been published in recent years [Mor73, Hal15, Ala08, Enr19, Huh12]. The Lagrangian trajectories of single tracer particles are used either to describe, for example the spreading of algal carpets in the oceans and weather ballons in the atmosphere (see Figure 2.18 i)), or to describe the conditions along a fluid particle trajectory in a stirred tank reactor (see Figure 2.18 ii)) [Ram01, Cam03, Kus20].

However, Lagrangian analyses have rarely been applied to the description of transport processes in process engineering apparatuses. An example application is the work of Llamas et al. or Kameke et al. [Lla20, Kam19], whereby the enormous potential of Lagrangian analyses as new and complementary methods to existing methods for characterization of process plants was shown.

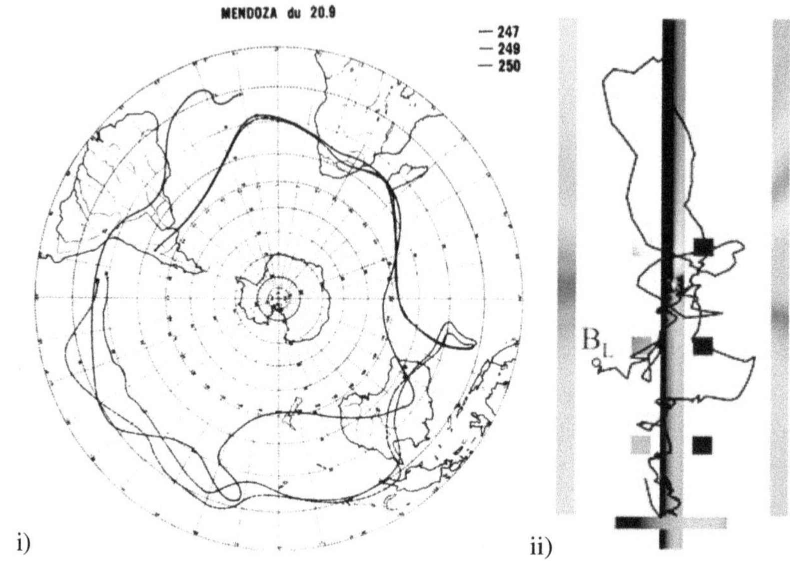

Figure 2.18. Example representation of trajectories: i) selection of three weather balloon trajectories related to the Eole experiment [Mor73]; ii) exemplary trajectory of a bacterium in an aerated stirred tank reactor [Kus20]

Furthermore in Kuschel et al. [Kus20], the authors have shown the normalized glucose and oxygen concentration experienced by an particular bacterium on a simulated trajectory in an aerated stirred tank reactor over a 200 s lifetime period. This is of particular interest not only due to the reactor geometry. Due to the oxygen and glucose specific consumption rates, different concentrations occur in different compartments, and thus a deficient supply of oxygen and glucose can occur for the bacterium.

For the Lagrangian description of particle motion, the position of the particle as a function of time must be known. Based on this information, the velocity as well as the acceleration of the particle can be determined as a function of the particle lifetime. In addition, the residence time of each particle in a specific region in the system can be directly specified on the basis of the temporal development of the particle position. Hereby, possible dead zones (areas of low mixing) or compartments in the system can be identified and described [Ram01].

Furthermore, the Lagrangian statistics of particle movements distinguishes between the single particle and the multi-particle view. In the following two subchapters the theoretical basics of the Lagrangian analysis methods with respect to residence time distribution and particle displacement were described.

2.2.1. Analysis of Particle Displacement

One of the most fundamental properties of Lagrangian particle motion is the absolute single-particle dispersion [HX13]. The absolute single particle dispersion

$$\langle \delta r^2 \rangle = \langle |\vec{r}(t) - \vec{r}(t=0)|^2 \rangle \tag{2.21}$$

describes the mean-squared displacement of a single particle moving along the trajectory $\vec{r}(t)$ with regard to its initial position $\vec{r}(t=0)$. In Figure 2.19 the absolute displacement of a single particle over time is plotted. The areas for which the absolute dispersion of the

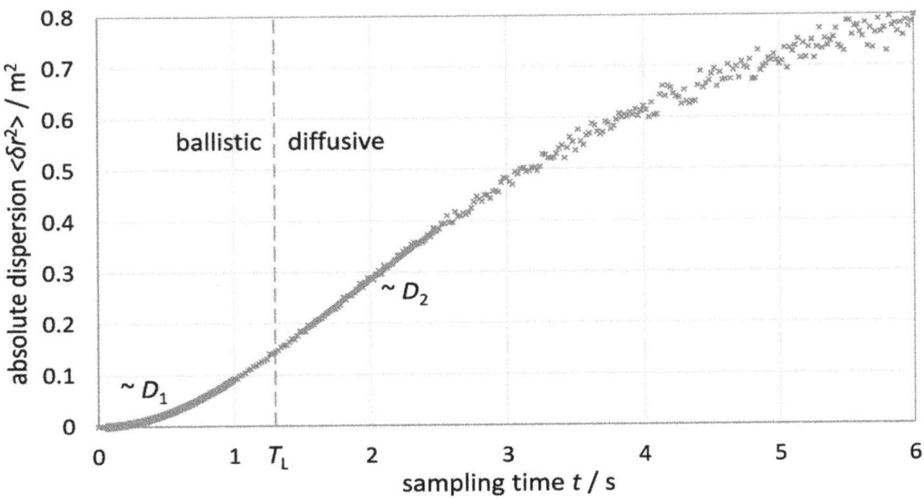

Figure 2.19. Visualization of the absolute dispersion of a single-particle over time; the two characteristic areas of ballistic and turbulent diffusive dispersion dependencies on observation time span (visualized in red), according to Xia et al. 2013 [HX13]

individual particles scale quadradically or linearly with time are highlighted (colored red line). For very large observation time spans, the course of the absolute dispersion flattens out again. This is due to the finite spatial extent which is given, for example, by the reactor geometry.

The typical single particle motion can be described according to EINSTEIN's theory of BROWNIAN motion, which assumes a ballistic motion for short observation times and a diffusive motion for long observation times [Tay21]. Thus, the following relationship can be observed for the absolute single particle displacement

$$\langle \delta r^2 \rangle \approx \langle u^2 \rangle t^2, \text{ at } t \ll T_\mathrm{L} \tag{2.22}$$

for short and

$$\langle \delta r^2 \rangle \approx \langle u^2 \rangle T_\mathrm{L} t, \text{ at } t \gg T_\mathrm{L} \tag{2.23}$$

long periods of observation [HX13], with the velocity u and the characteristic Lagrangian integration time T_L. For short observation periods the single particle displacement depends on the time squared, and for long observation periods it depends linearly on the time. The Lagrangian integration time

$$T_\mathrm{L} = \int F(t)t. \tag{2.24}$$

depends on the Lagrangian autocorrelation function

$$F(t) = \frac{\langle u(t_0 + t)u(t_0) \rangle}{\tilde{u}^2} \tag{2.25}$$

with the particle velocity $u(t)$. Based on these assumptions, the turbulent diffusion coefficient

$$D_2 = \langle u^2 \rangle T_\mathrm{L} \tag{2.26}$$

can be calculated for large time periods. For this, however, the Lagrange autocorrelation function must be calculated, since this cannot be predicted theoretically so far. Furthermore, this cannot be predicted or calculated from Eulerian velocity data [HX13]. Additionally, the ballistic dispersion coefficient D_1 can be calculated by polynomial fitting of the dispersion data for short time periods according to Eq. 2.22.

In the case that the single particle approaches or even returns to its point of origin, the value for the absolute dispersion will be zero again. Due to the definition of absolute dispersion, negative values are not possible. In addition to the absolute single particle displacement, the interaction between two particles is often considered. Here, the displacement of the particles is considered in relation to their center of mass, which corresponds to the displacement of a particle pair over time. For this, those particle pairs are considered, which have met at the same time at a certain place [Bak08].

Due to the fact that experimentally no two particles can be added to the system at the same place and time, a minimum particle-particle distance ε must be defined, so that two particles can be assumed as a particle pair. The larger the allowed distance at the beginning of the observation, the higher the probability that the velocity vectors of the two particles do not correlate with each other. For this reason, the relative dispersion is strongly dependent on the particle-particle distance ε and thus less meaningful than in the case where two particles are added to the system at the same place. For this reason, only the absolute dispersion is considered below.

2.2.2. Sojourn Time Distribution

Another quantity that can be calculated directly from the Lagrangian trajectories is the local residence time of each particle for a defined subvolume. To avoid confusion of the definition of the Lagrangian residence time of particles with the conventional definition of the hydraulic residence time in continuous flow systems, the residence time of the Lagrangian trajectories is referred as the Sojourn time [Ram01].

Depending on the size of the particles, the subvolume and the type of flow, the local residence time is directly proportional to the reciprocal of the local velocity. In general, the greater the local velocity of the particle, the lower the local residence time. However, the local residence time deviates from the reciprocal of the velocity if the particle has a high velocity but circulates within the subvolume.

The first experimental determination of the Sojourn time distribution for individual compartments in a stirred tank reactor was shown by Rammohan et al. [Ram01]. For this purpose, the trajectory of a radioactive gamma source in a stirred tank reactor was reconstructed with the help of a gamma ray detector matrix. For a sufficient analysis, the single particle in the system had to be observed over a very long period of time.

Based on these findings, several Lagrangian particles were simulated in a stirred tank reactor and the local Sojourn time was calculated in a further numerical work. Figure 2.20 shows an example of such a local Sojourn time distribution from Campolo et al. [Cam03]. The Sojourn time is visualized by the three-dimensional structure and the height of the surface of this structure. The special feature of this visualization is that, based on the Sojourn time distribution, zones of good or poor mixing can be determined for the system under investigation. This is applicable on the one hand for the characterization of stirred tank reactors and on the other hand for a better understanding of the transport processes in

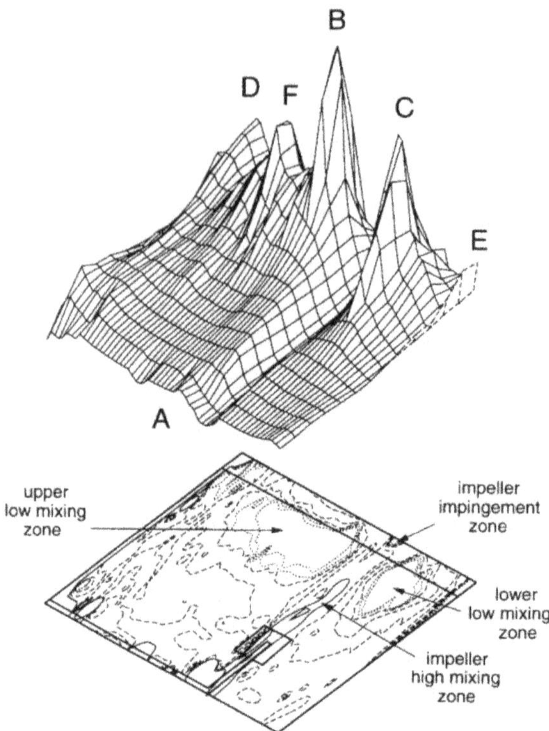

Figure 2.20. Visualization of the projection of the local Sojourn time distribution in a single-stage stirred tank reactor based on numerical Lagrangian particles; A-E different zones in the reactor [Cam03]

the system. Based on these results, it is also possible to design new processes more efficiently.

Furthermore, the resolution as well as the magnitude of the local Sojourn time depends on the temporal and spatial resolution of the particle tracking measurement system used as well as the subsequent definition of the size of the subvolumes. Due to this, an indication of the spatial and temporal resolutions is mandatory for visualizing the Sojourn time in order to put the results in context with other data.

2.3. Conclusion

Based on the described fundamentals of power input and mixing time of aerated stirred tank reactors, it can be concluded that a large number of papers on these topics can already be found in the literature. Nevertheless, most of the work is based on integral or global investigations. More detailed insights into the transport processes of aerated stirred tank reactors are rare or based on two-dimensional flow field measurements.

Furthermore, it is evident that the main characterization methods are based on Eulerian approaches. Modern approaches for the description of transport processes in aerated stirred tank reactors, which are based on Lagrangian approaches, have only been applied in a few exceptional cases.

With the described methods, significantly deeper insights into the transport processes in stirred tank reactors can be generated. For example, the method for determining the mixing time not only allows the global mixing time to be determined, but also the history of the mixing to be visualized. Based on this, it is also possible to identify coherent structures such as compartments or dead zones. Furthermore, the 4DPTV method allows the Lagrangian description of aerated stirred tank reactors based on Lagrangian particle trajectories. The conditions along a trajectory can be statistically evaluated. In this work, the methods for describing local phenomena as well as Lagrangian methods are used to describe the local as well as global transport processes in aerated stirred tank reactors.

3. Experimental Design and Procedure

In the following, the respective experimental setups for the investigations of the local mixing time distribution and the measurement of the Lagrangian trajectories are described. Since both investigations were carried out on the same stirred tank reactor, the experimental setup is first described in general. Next, possible additions to the experimental setup are discussed in the respective subchapters. The section on the experimental setup is followed by a detailed description of the individual experimental procedures.

3.1. 3 L Lab Scale Stirred Tank Reactor Setup

In the following, the experimental setup for determining the local mixing time as well as the Lagrangian trajectories is described in detail. For these studies, a copy of a commercially available 3 L glass stirred tank reactor (*3 L-single wall glass autoclavable bioreactor*, Applikon, The Netherlands) is used, as this does not have any disturbing imprints on the outer wall of the reactor (labeling, level indicator, brand logo). A schematic visualization of the stirred tank reactor and a flow diagram are shown in Figure 3.1. A photograph of the experimental setup is shown in Figure 3.2.
The inner diameter of the stirred tank reactor is $D = 130$ mm. The reactor is equipped with either two standard Rushton stirrers (Figure 3.1 f)) with a diameter of $d_{\text{stirrer}} = 36$ mm or three elephant ear stirrers ($d_{\text{stirrer}} = 60$ mm). The bottom distance s of the first stirrer is $s = 1.5d$ in both cases, and the distance between the two stirrers is $s = 2.2d$ respectively $s = 1d$ for the tripple Elephant Ear configuration. The working volume is $V_{\text{fill}} = 2.8$ L, which corresponds to an aspect ratio of $H/D = 1.83$. The bottom geometry of the reactor is a torispherical bottom according to DIN 28013. For better optical access, the stirred tank reactor is surrounded by an octahedral basin with an edge length of 176 mm and a height of 270 mm. For optical adjustment of the refractive index, a water-glycerol mixture of 98 % glycerol with a refractive index of $\lambda = 1.47$ was adjusted in the outer basin.

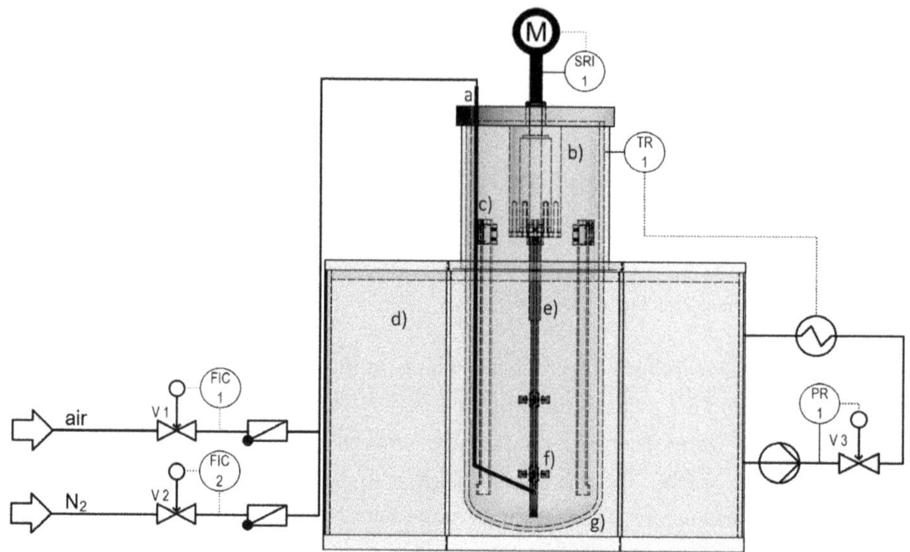

Figure 3.1. Representation of the stirred tank setup : a) sparger, b) bearing box, c) baffle and baffle holder, d) octahedral water-glycerol basin, e) stirrer shaft, f) Rushton turbine, g) bottom geometry

All experiments are performed with DI-water at ambient temperature of $T = (25 \pm 1)^\circ$C. The temperature in the stirred tank reactor is controlled by an external water-glycerol basin (Figure 3.1 d)). The outer water basin is connected to an external heating circuit.

For two-phase investigations (water/air), the reactor system is aerated with compressed air via an open tube sparger. The gas outlet nozzle of the sparger used is $d_{\text{nozzle}} = 2$ mm. The sparger (Figure 3.1 a)) is located 10 mm below the lower stirrer at a radius of 15 mm. The nozzle thereby faces tangentially in the direction of stirrer rotation. The gas volume flow can be controlled via a gas mixing unit. Two volume flow controllers (*EL-Flow Select F-200CV*, Bronkhorst, The Netherlands) are used for this purpose; each is calibrated for a specific gas or gas mixture. Based on the recommendations in Zlokarnik 2001 [Zlo01], the reactor is equipped with three baffles made of acrylic glass (Figure 3.1 c)). The stirrer shaft is connected to the motor via a bearing box (Figure 3.2 b)) on the top of the reactor. The stirrer speed is regulated and controlled with a laboratory stirrer (*ViscoPakt-X7*, HiTec Zang, Germany).

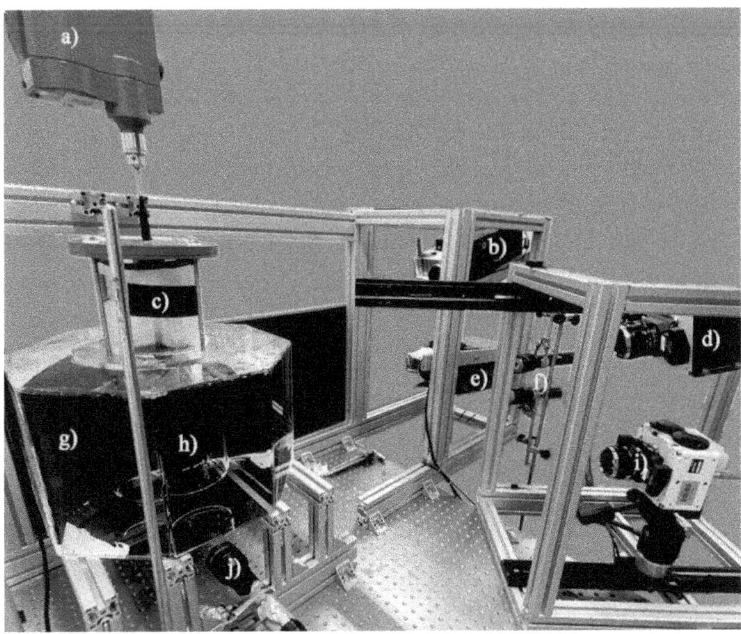

Figure 3.2. Image of the experimental setup : a) motor, b) 1. *hs.dimax HS2*, c) stirred tank reactor, d) 1. *Phantom VEO640s*, e) 2. *hs.dimax HS2*, f) 1. & 2. high power LEDs, g) octahedral water-glycerol basin, h) measured volume, i) 2. *Phantom VEO640s*, j) 3. high power LED

The setup in Figure 3.1 corresponds to the principal setup which can be used for the mixing time studies and the measurements of the Lagrangian trajectories. Possible adaptations to the respective experiments or extensions of the basic setup are described in the corresponding Chapters.

Table 3.1. Overview of the general process parameters

Parameters	**Settings**
Temperature T / °C	25
Reactor Volume V / L	2.8
Stirrer frequency n / s^{-1}	150, 450
Aeration q / vvm	0 & 0.06

3.2. Local Mixing Time Determination

The method for determining and evaluating the local mixing time distribution has already been described in detail in Fitschen et al. [Fit21].

For the determination of the local and global mixing time by means of the decolorization method, the time course of an acid/base neutralization reaction is observed. Bromothymol blue is used as the pH indicator for this purpose. In the following, the experimental setup specific for the mixing time determination is described. For better understanding, Figure 3.3 shows an example of the setup from a top view. A two molar sodium hydroxide (CAS: 1310-73-2) solution and a two molar hydrochloric acid (CAS: 7647-01-0) solution are used for the experiments. 200 µL of a 10 % bromothymolblue-ethanol (CAS: 76-59-5 and 64-17-5) solution are used for the experiments. The diluted acid that induces the color change is added

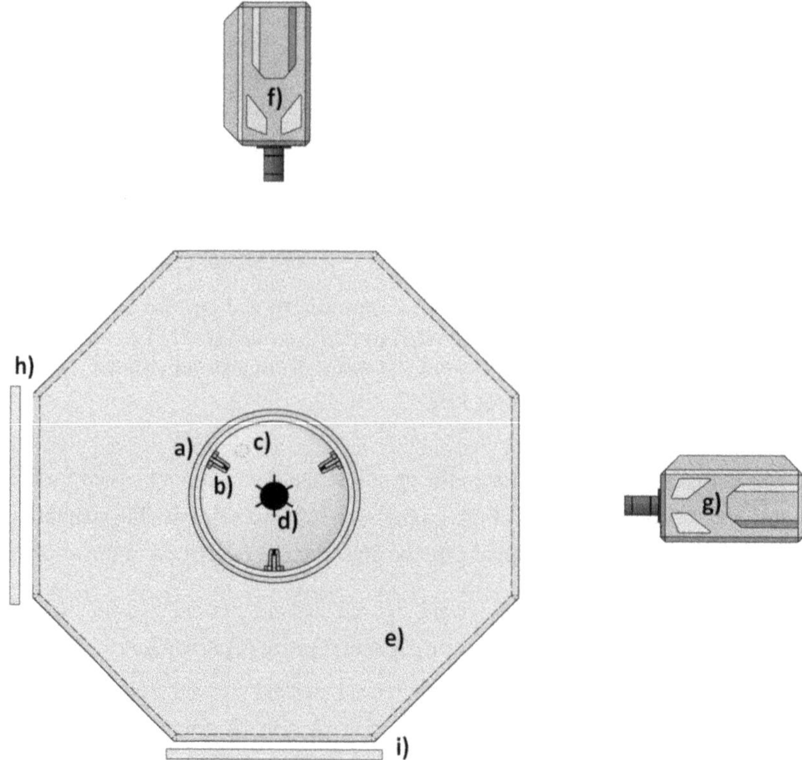

Figure 3.3. Representation of the stirred tank reactor setup: a) stirred tank reactor, b) baffle and baffle holder, c) point of addition, d) stirrer, e) octahedral water basin, f) & g) camera, h) & i) LED-panel

using a syringe pump (*type 11 Elite Infusion Only*, Harvard Apparatus, USA). The point of addition (Figure 3.3 c)) is located at a radius of $r_A = 50$ mm and an angle of 18° to the baffle. The addition was made 1 cm below the water surface. The diluted hydrochloric acid is added to the system via a 0.8 mm cannula and a nozzle outlet speed of 0.0895 m/s.

Two cameras (*D7500*, Nikon, Japan), each with a 50 mm fixed focal length lens (*Makro-Planar T* 2/50 mm*, Zeiss, Germany), were used to record the color change (Figure 3.3 f & g)). Here, the two cameras are arranged at an angle of 90° to each other in order to be able to view the temporal course of the mixing from two perspectives. The cameras are mounted at a distance of 800 mm from the center of the agitator shaft, with the focal plane on the agitator shaft. The exposure time for both cameras is $f = 1/400\,\text{s}^{-1}$ with an aperture of $A = 4.5$. The videos are recorded with a local resolution of 1920 x 1080 px^2 and a temporal resolution of 60 Hz.

Since bromothymol blue absorbs different wavelengths of light depending on the pH in the system, a warm white light from an LED panel is used as the reference signal. One LED panel (Figure 3.3 h & i)) per camera is used on the opposite side of the reactor from the camera. To synchronize both camera signals to each other, the backlight is switched off at the beginning of each measurement. The background illumination is then switched on at the same time the diluted hydrochloric acid solution is added. As a result, an intensity jump can be seen in each camera signal at the time of addition.

Table 3.2. Overview of the equipment used and the test parameters for determining the mixing time

Parameters	**Settings**
Camera	2x *Nikon* D7500
Objective	*Nikon* Makro-Planar T* 2/50 mm
f-number	4
Illumination	2x white constant current LED panels
Tracer	bromothymolblue
Frame Rate	60 Hz
Acquisition time	various
Spatial resolution (raw)	150 µm
Spatial resolution (Eulerian)	150 µm
Software	self written code

3.3. 4D Particle Tracking

In the following, the modifications to the experimental setup necessary for the determination of the Lagrangian trajectories as well as the additional measuring equipment required for this purpose are described. Afterwards, the experimental procedure is described in detail. Figure 3.4 shows the arrangement of the cameras to the stirred tank reactor and the calibration target used. Four cameras (two *hs.dimax HS2*, PCO, Germany, and two *Phantom VEO640s*, Vision Research Inc, USA) are installed at a distance of 80 cm from the stirrer shaft. Two identical models are installed respectively one above the other at an angle of 30°. The two camera pairs are mounted at an angle of 45° to each other (see Figure 3.4 i)). All cameras are equipped with a $f = 50$ mm fixed focal length lens (*Makro-Planar T* 2/50 mm*, Zeiss, Germany).

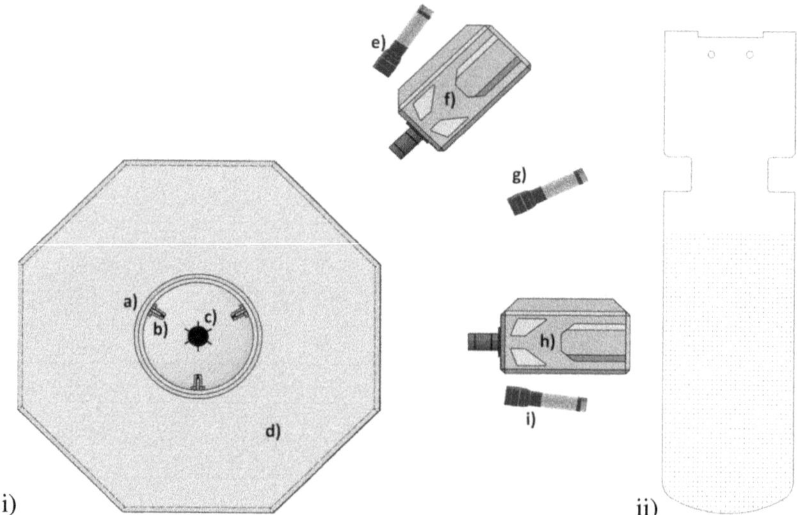

Figure 3.4. Representation (i) top view (1:10), i) target) of the stirred tank setup : a) stirred tank reactor, b) baffle and baffle holder, c) stirrer, d) octahedral water basin, e, g & i) high power led light, f & h) camera

In addition, all cameras are equipped with a Scheimpflug adapter to correct the focus plane of all cameras to the same plane. For the calibration of the cameras, a target with a dot matrix (dot size: 1 mm, distance to each other: 5 mm) is used. The target has a slightly smaller contour than the inner geometry of the stirred tank, so that the target can be shifted by $\Delta x = \pm 20$ mm around the center axis parallel to the focal plane.

Fluorescent orange polyethylene microspheres with a density of $\rho = 1.0$ g/ml and a diameter of $d_P = 0.150 - 0.180$ mm are used as tracer particles. For the volume illumination of the particles in the stirred tank reactor, three high-power LEDs (*LE B P3W 01* from *Osram OSTAR*, Germany) are used, which are controlled by means of self-developed high-power electronics and have each a pulsed power of about $P_{LED} = 40$ mJ. Two light sources shine from the direction of the two camera pairs, and one light source shines from below the tank. To prevent the light of the high-power LEDs from interfering with the fluorescence signal of the particles, a high-precision OD 4 long-pass filter (*TECHSPEC*, Germany) with a cut-off frequency of 575 nm is installed in front of each camera chip.

The high-speed recordings were controlled by the software *DaVis* from *LaVision* and synchronized with the corresponding PTU (Programmable Timing Unit) from *LaVision*.

Table 3.3. Overview of the parameters for the 4DPTV measurements and the equipment used

Parameters	**Settings**
Camera	2x *hs.dimax HS2* and 2 x *Phantom VEO640s*
Objective	*Nikon* Makro-Planar T* 2/50 mm
f-number	8
Illumination	3x blue high power LEDs *LE B P3W 01* from *Osram OSTAR*
Seeding Particles	*Cospheric* fluorescent orange polyethylene microspheres 1.00 g/cm^3 and 10 µm
Frame Rate	800 Hz
Acquisition time	3.75 s
Spatial resolution (raw)	154 µm
Spatial resolution (Lagrange)	154 µm
Spatial resolution (Eulerian)	1.21 mm (time-averaged), 1.5 mm (instantaneous)
Software	*DaVis* 10.2 form *LaVision*

4D Particle Tracking and Measurement Procedure

The following describes the experimental procedure and methods used for the Lagrangian particle tracking. First, the recording and post-processing of the images will be discussed, followed by the evaluation routine for particle tracking. The algorithm of Schanz et al. [DS16] is used for particle tracking. This method has the decisive advantage over other particle tracking methods in that it is able to triangulate particles in space reliably and efficiently and to generate almost no ghost particles [DS16].

All measurements shown in this thesis are performed with a temporal resolution of 800 Hz. The exposure time of the cameras as well as the illumination time of the LEDs is 800 μs, with the signal from the LEDs delayed by 10 μs to ensure that all camera shutters are fully open. Each set of measurements involves first taking images for the purpose of geometric calibration and volume selfcalibration. Thereafter, 3000 images are recorded for each experiment. The number of images is limited to 3000 by the internal memory of the cameras.

For the geometric calibration, various shifted images are taken of the target shown in Figure 3.4. Subsequently, a geometric model can be created on the basis of the target images, which assigns the same coordinate system to each camera and compensates for optical differences. Due to the fact that the smallest changes to the setup lead to a decrease in the geometric model quality, the recordings of the target as well as the software calibration must be carried out before each series of measurements.

A measurement at reduced seeding density is used for volume self-calibration, which is done with the commercial *DaVis 10.2* software to improve the geometric model to a voxel error of less than 0.1 voxel. For the final 4D particle tracking velocimetry measurements, a seeding density of 0.05 ppp (particle per pixel) is used.

Due to the fact that two different camera types are used in the setup, an additional post-processing step is implemented to adjust the respective intensity signals of the individual particles in the image compared to the background noise for the two camera types. This is necessary because particle triangulation does not work reliably when the signal intensities of the two camera types are different from each other. Figure 3.5 shows a section of a captured raw (i)) and processed images (ii)) to illustrate the editing process. The post-processing algorithm is first used to subtract a time-averaged minimum image from all raw images (Figure 3.5 i)). Then, a second-order Butterworth filter [Ell12] is used to improve the signal-to-noise ratio. In addition, sharp gradients in the intensity signal of the images are smoothed. Furthermore,

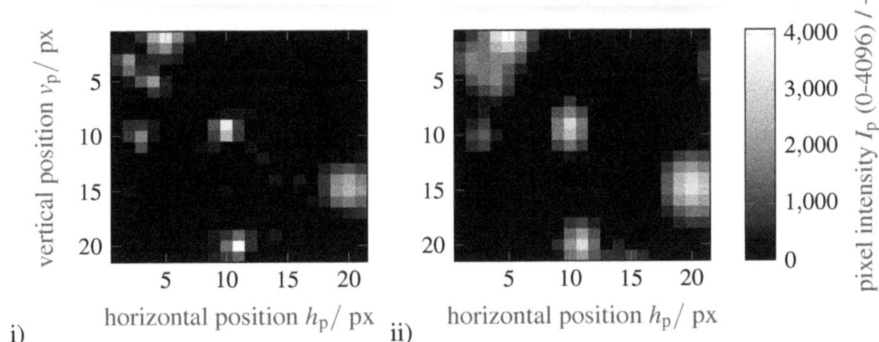

Figure 3.5. Illustration of captured and processed particle image (i)) raw captured image (ii)) filtered image

the Butterworth filter reduces the risk of peak-locking which leads to a bias-error in particle detection [Mic16]. Based on this procedure, the particles are more clearly highlighted from the background, which improves the particle detection (Figure 3.5 ii)).

The parameters set for particle tracking with the shake-the-box method and the parameters for the subsequent calculation of the Eulerian temporally resolved and averaged velocity field are listed in Table 3.4 & Table 3.5.

Table 3.4. Overview of the settings for the Shake-The-Box algorithm in *DaVis* 10.2

Parameters	**Settings**
Threshold for 2D particle detection / counts	250
Allowed triangulation error / voxel	1.5
Adding particles (outer loop)	10
Refine particle position (inner loop)	16
Shake particle position by / voxel	0.6
Remove particle if closer than / voxel	0.25
Make Optical Transfer Function (OTF) smaller by	0.85
Residuum computation by	1.1
Software	*DaVis* 10.2 form *LaVision*

Table 3.5. Overview of the settings for the calculation of the time resolved and averaged 3D velocity field in *DaVis* 10.2

Parameters	**Settings**	
compute:	average flow field	transient flow field
subvolume size/ voxel	32	72
overlap / %	75	87.5
min. number of particles	2	2
spatial polynominal order	3	3
track fitting polynominal order	2	2
filter length (time steps)	2	2
min. track length (time steps)	4	4
Software	*DaVis* 10.2 form *LaVision*	

4. Experimental Results and Discussion

The results of this thesis are summarized and discussed in the following. First, the results of the hydrodynamic characterization of the stirred tank reactor based on the velocity fields and the power input are presented. Afterwards, the results of the local mixing time characterization are discussed. Finally, Lagrangian analysis of the 4DPTV measurements in a stirred tank reactor are presented and discussed. The results are followed by a summary evaluation and discussion of the results.

4.1. Hydrodynamic Characterization

The velocity vector fields shown in the following are based on the 4DPTV measurements and a subsequent conversion of the Lagrangian particle trajectories to an Eulerian grid. As with the selection of the interogation area in PIV (Particle Image Velocimetry) for the calculation of the vector field, a subvolume is defined for the projection of the Lagrangian data onto the Eulerian grid in which the data of the particle trajectories are averaged. For the evaluation, a local distribution of 32 voxels with an overlap of 75 % was chosen, resulting in a grid spacing of $\delta x = 1.2194$ mm.

However, due to the arrangement of the cameras and the resulting two horizontal perspectives (see Figure 3.4 & Figure 4.1), no particle trajectories could be determined on the side of the stirrer shaft and stirrer elements facing away from the cameras and thus, no Eulerian vector fields can be calculated. Furthermore, artifacts are created around the areas that cannot be evaluated due to the optical impairment. For this reason, only the front third of the reactor between two baffles is considered for the following illustrations and evaluations (see Figure 4.1 dashed red line). Additionally, the area of the bottom geometry is also excluded from the evaluation due to the fact that no satisfactory geometric calibration is currently possible here due to strong optical degradation. Figure 4.2 and Figure 4.3 show the time-averaged velocity fields for the two reactor geometries (2x Rushton stirrer and 3x

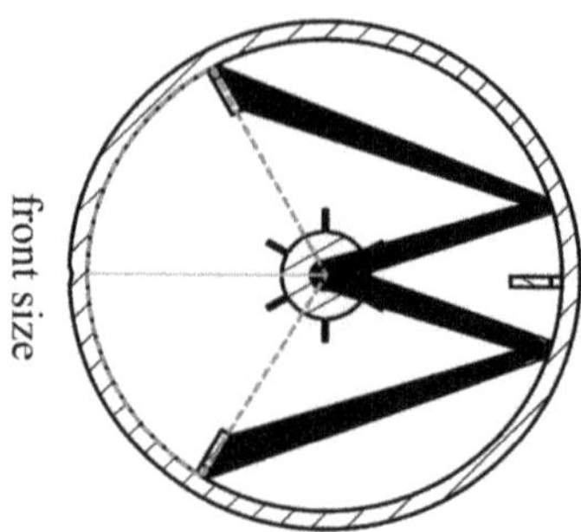

Figure 4.1. Top view representation of the volumes in the stirred tank reactor that cannot be evaluated due to the optical barriers; red: evaluable volume; green: slice plane for vector field evaluation; black: shading due to internals (baffles, stirrer shaft)

Elephant Ear stirrers) and for three different stirrer frequencies. For the representation of the

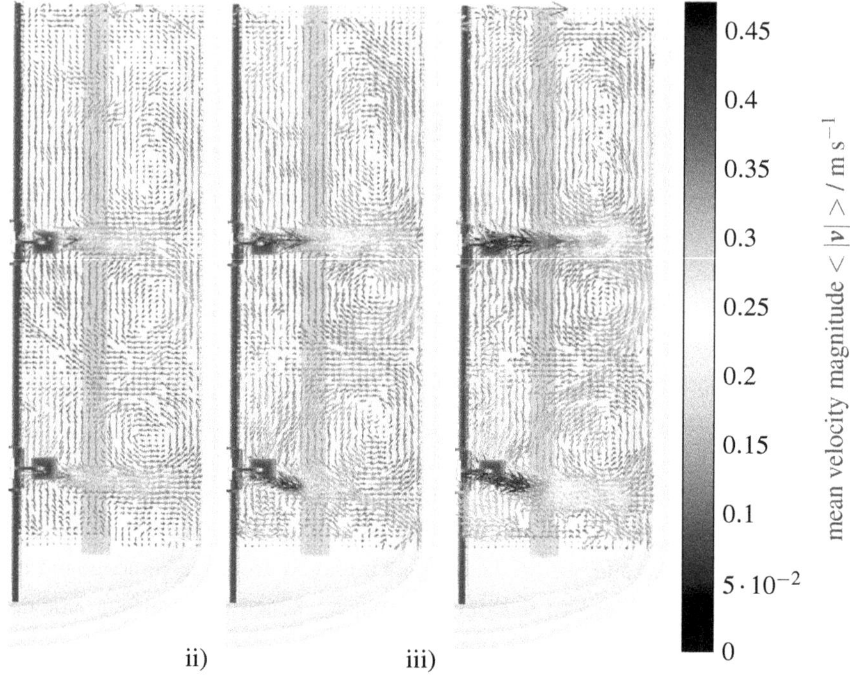

Figure 4.2. Three dimensional and time-averaged velocity field for the double Rushton turbine configuration; slice position referred to Figure 4.1; i) $n = 250$ rpm, ii) $n = 350$ rpm, iii) $n = 450$ rpm

velocity vector fields, a uniform color scale was chosen so that the similarities and differences in the structures can be better compared. The interaction of the two Rushton stirrers can be seen in the illustrations in Figures 4.2 i - iii). Starting from the radial discharge of the two stirrers, the typical vortices caused by the deflection of the flow at the reactor wall can be seen. In addition, it can be clearly seen that regardless of the rotational frequency, the upper stirrer causes a horizontal outward flow. The lower turbine has a slightly downward flow angle. The lower Rushton turbine causes a downward flow due to the bottom geometry of the stirred tank reactor. Compared to a cylindrical flat reactor bottom, the influence of the baffles does not extend into the dished bottom area. Because the dished bottom is not equipped with baffles, a circulating flow is formed in the reactor bottom, which is generated by the lower stirrer. Figures 4.3 i - iii) illustrate the Eulerian velocity fields of the 4DPTV measurements for the triple Elephant Ear stirrer geometry. Figures 4.3 i - iii) demonstrate the enormous

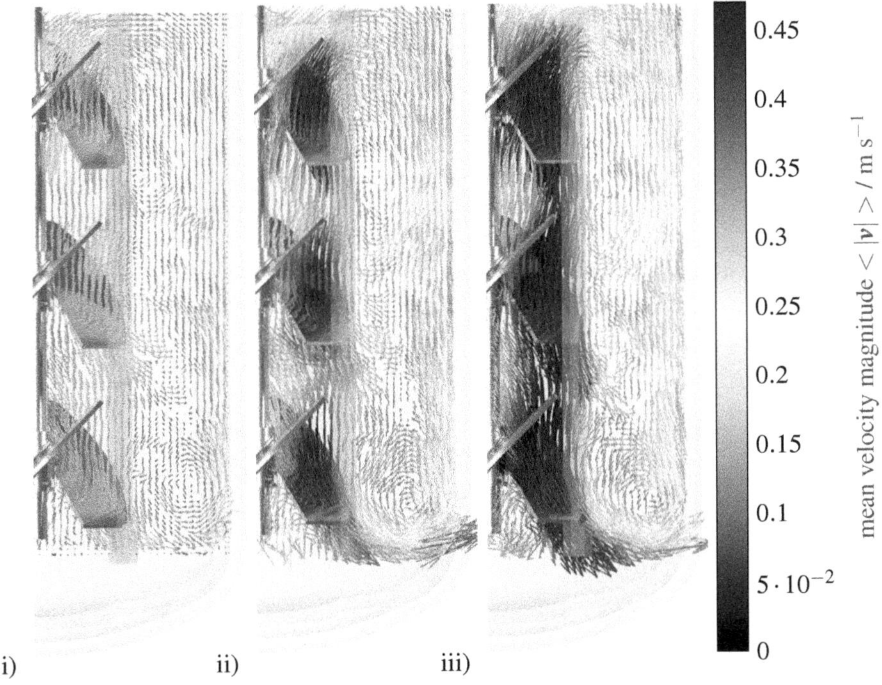

Figure 4.3. Three dimensional and time-averaged velocity field for the triple Elephant Ear stirrer configuration; slice position referred to Figure 4.1; i) $n = 150\,\text{rpm}$, ii) $n = 250\,\text{rpm}$, iii) $n = 350\,\text{rpm}$

possibilities offered by the three-dimensional tracking of the particle trajectories because even in the areas swept by the large stirrer blades it is possible to resolve the velocity field using the 4DPTV measurement. In particular, the use of high-power LEDs to illuminate the volume means that every particle in the system can be induced to fluoresce and can therefore be detected by the respective cameras. Furthermore, Figures 4.3 i - iii) demonstrate the axial character of the Elephant Ear stirrers, which generates a downward flow close to the stirrer shaft. The volume displaced downward by the stirrers is deflected in the bottom geometry and transported back upward near the wall. The reactor geometry can thus be divided into two rings, the inner ring in which the volume is conveyed downward by the stirrers and the outer ring in which the volume rises upward again near the wall. Due to the significantly larger volume in the outer ring, the average flow velocity there is significantly lower. The interaction of these two opposing flows also causes local vortices in the transition region between the two large-scale flows. This effect can especially be observed in the area of the lowest stirrer.

Due to the size of the Elephant Ear stirrers, their actively stirred volume is significantly larger compared to the Rushton stirrers. This can be seen in particular in Figures 4.4 & 4.5. Figures 4.4 and 4.5 illustrate the relative frequency distributions for the respective stirrer

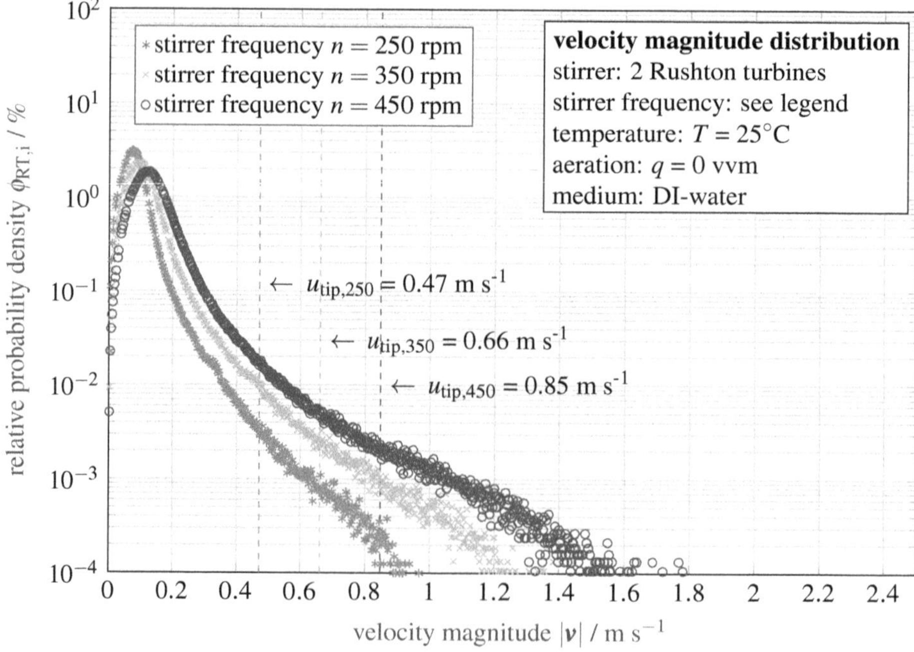

Figure 4.4. Comparison of the relative frequency distribution of the velocity magnitude for different stirrer frequencies ($n = [250, 350, 450]$ rpm and no aeration) in the 3L reactor; double Rushton turbine configuration; class width: 10 bit resolution

combinations and stirrer frequencies for a 10-bit division of the x-axis. The probability frequency distributions are based on the Lagrangian velocity data, since a conversion to the Eulerian grid does not make sense. This is because an interpolation of the Lagrangian data onto an Eulerian grid would lead to a loss of information. Further, the projection of the Lagrangian data onto the Eulerian grid is computationally intensive and leads to a compression of the data set. From Figure 4.5 it can be seen that the probability of the most

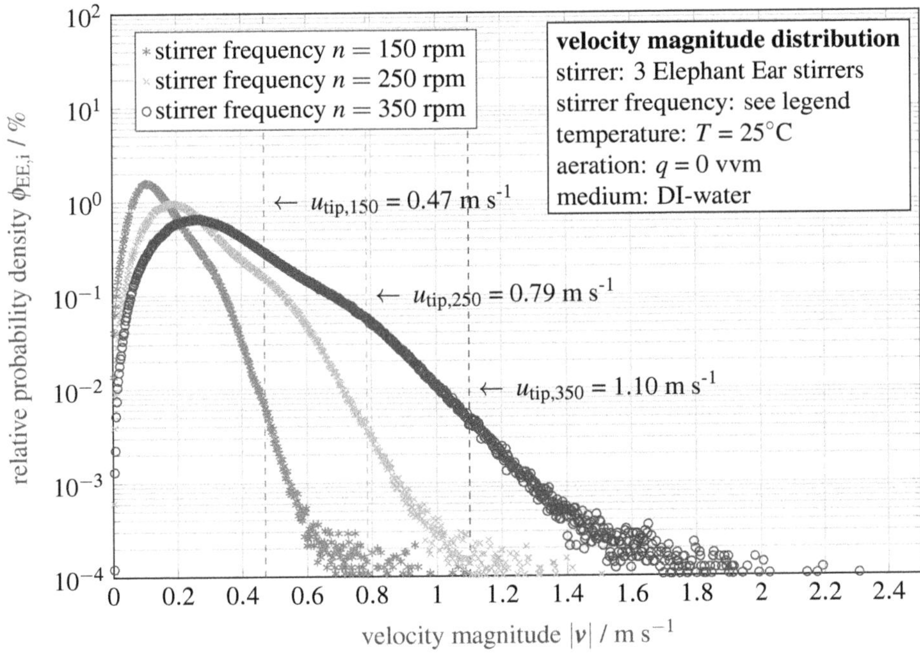

Figure 4.5. Comparison of the relative probability frequency distribution of the velocity magnitude for different stirrer frequencies ($n = [150, 250, 350]$ rpm and no aeration) in the 3L reactor; tripple Elephant Ear configuration; class width: 10 bit resolution

common velocity is lower for the triple Elephant Ear stirrer geometry than for the double Rushton turbine combination. This is due to the fact that, the velocity density distribution for the triple Elephant Ear stirrer configuration is much broader and thus necessarily flatter, since the integral of the distribution must yield 100 %.

Furthermore, in case of the same stirrer tip speed of $u_{tip} = 0.47$ m/s, the maximum measured particle velocity magnitude for both stirrer configurations is approximately $|v| \approx 1$ m/s.

It should be noted that it is not impossible that velocities greater than the stirrer tip speed may occur. This effect occurs especially in the trailing vortices behind the stirrer blades [Van75]. Moreover, probabilities of higher velocities are significantly lower for the triple Elephant Ear stirrer configuration. This is interesting since the actively stirred volume is significantly larger for the triple Elephant Ear stirrer configuration and thus a higher probability of higher velocities would have been expected.

In addition to the velocity probability frequency distributions, the probability frequency distribution can also be visualized for the resulting accelerations. The measured accelerations in the system are directly proportional to the forces acting on the particles. The acceleration probability frequency distributions are shown in Figures 4.6 and 4.7 for a 10-bit division of the x-axis. Analogous to the velocity probability frequency distribution, the distribution for

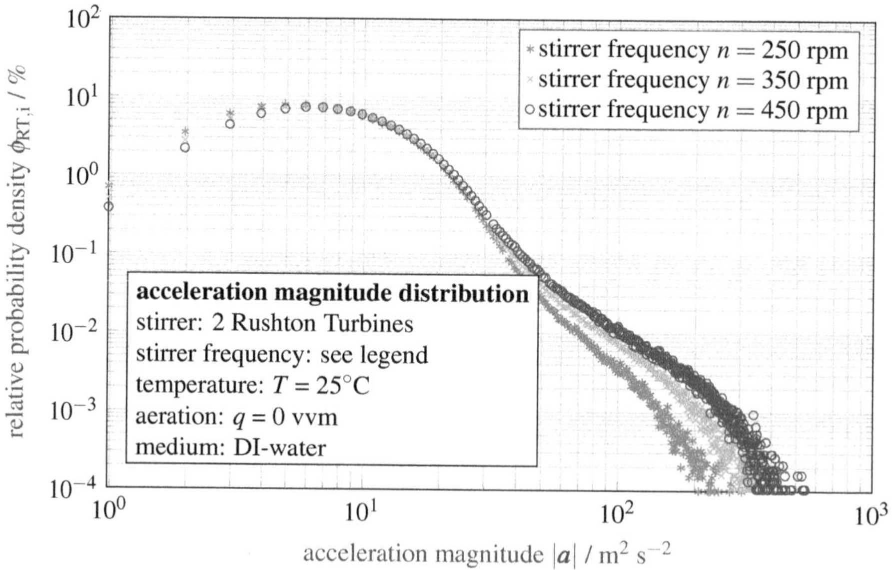

Figure 4.6. Comparison of the relative probability frequency distribution of the velocity magnitude for different stirrer frequencies ($n = [250, 350, 450]$ rpm and no aeration) in the 3L reactor; double Rushton turbine configuration; class width: 10 bit resolution

the acceleration in the system with three Elephant Ear stirrers is slightly flatter and therefore broader than in the system with two Rushton stirrers. Furthermore, the maximum measured accelerations in the system with three Elephant Ear stirrers are slightly higher. However, based on the data in Figures 4.4 and 4.5 it was shown that the maximum velocities occurring in the two systems are very similar. The acceleration data shown are extracted directly from

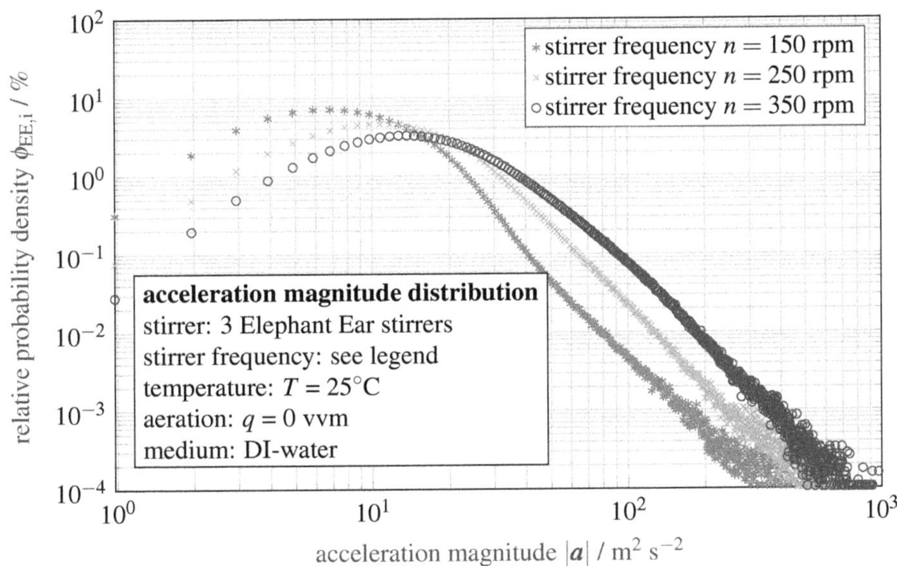

Figure 4.7. Comparison of the relative probability frequency distribution of the acceleration magnitude for different stirrer frequencies ($n = [150, 250, 350]$ rpm and no aeration) in the 3L reactor; tripple Elephant Ear configuration; class width: 10 bit resolution

the Lagrangian data sets and do not have anything to do with the velocity fluctuations of the Eulerian approach and thus cannot be used to calculate the local energy dissipation. A analysis of the local energy densities and dissipation rates is given in the following chapter.

4.1.1. Characterization of Stirred Tank Reactors by Means of the Volumetric Power Input

To describe the power input characteristics of stirred tank reactors, the power number *Po* is visualized as a function of the Reynolds number *Re*. For the stirrer geometries used in this work, the dependence of the power number on the Reynolds number is plotted in Figure 4.8 based on the torque measurements. The range of Reynolds numbers shown in Figure 4.8 cannot be reproduced by varying the stirrer frequency alone, due to the fact that the resulting torques would no longer be measurable at very low stirrer frequencies in a aqueous medium. Therefore, the viscosity is also varied by means of a water-glycerin mixture.

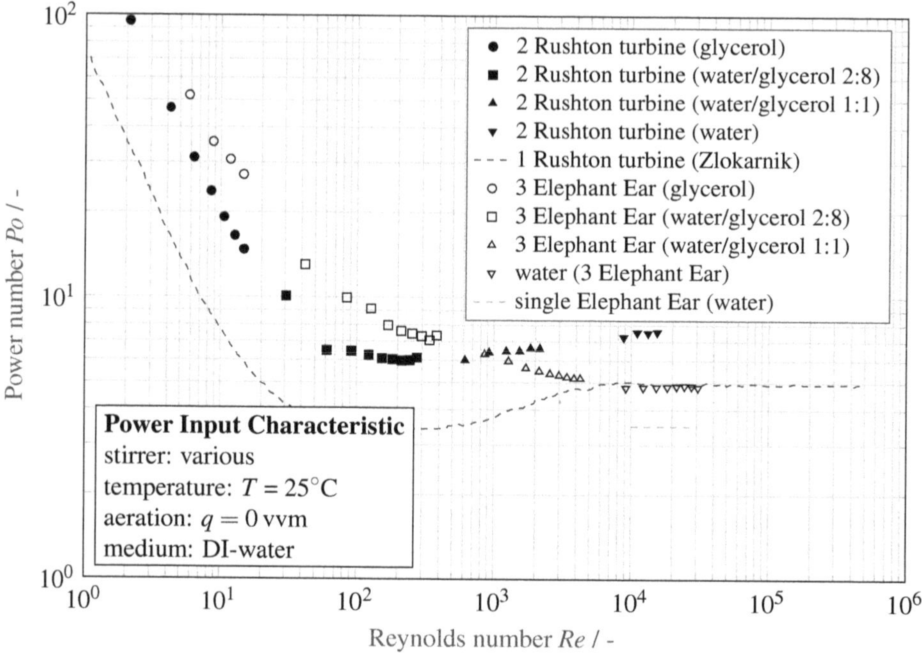

Figure 4.8. Global power input measurements: Visualization of the power number *Po* as a function of the Reynolds number *Re* for different stirrer types

A comparison with the literature data of a simple Rushton turbine according to Zlokarnik [Zlo01] shows very good agreement of the dependence of the power number *Po* on the Reynolds number *Re*. The trend for the double Rushton turbine combination is only scaled towards larger values. However, it can also be observed that in the turbulent range ($Re > 10^4$)

the use of two Rushton stirrers does not lead to a doubling of the power input Po. This can be explained by the fact that the two stirrers interact with each other and are therefore not independent. For the double Rushton turbine configuration in the turbulent region, the power number equals $Po = 7.5 \pm 0.2$.

Additionally, Figure 4.8 shows the power input characteristics for the configuration with the three Elephant Ear stirrers. In the laminar flow region, the three Elephant Ear stirrers have a significantly higher power number than the double Rushton configuration. This trend reverses in the turbulent flow region, and the power number drops to $Po = 4.8 \pm 0.1$.

A more detailed description of the power input into the system can be provided by analyzing the energy cascade in the three Cartesian space directions. Therefore, the calculation of the absolute kinetic energy is shown and discussed in the following, using the double Rushton configuration as an example. In Figure 4.9 the time-averaged total kinetic energy E is plotted as a function of the wave number k for the two-stage Rushton configuration.

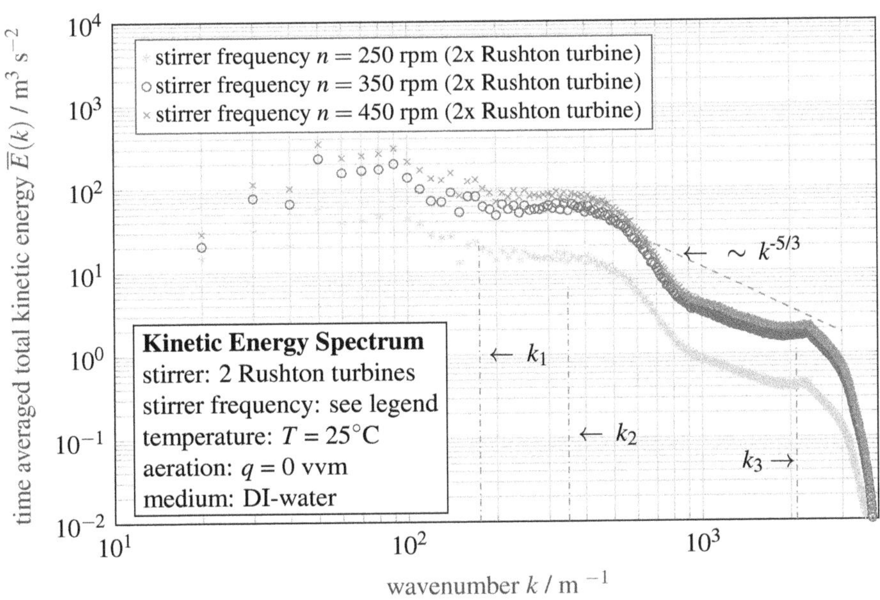

Figure 4.9. Energy cascade for the double Rushton turbine configuration for different stirrer frequencies n; (k_1^{-1}:stirrer diameter, k_2^{-1}: stirrer radius and k_3^{-1}: stirrer blade height)

The calculated absolute kinetic energy spectra are based on the instantaneous velocity fields which were averaged over time. In addition, the slope of $-5/3$ theoretically derived by KOLMOGOROV is plotted on Figure 4.9. Additionally, the typical lengths (k_1^{-1}:stirrer diameter, k_2^{-1}: stirrer radius and k_3^{-1}: stirrer blade height) for the two stage Rushton turbine combination are plotted in Figure 4.9, since the energy is introduced into the system on these scales.

The wave number range shown here is limited by two characteristic values. First, the wave number is limited downward by the maximum dimension of the investigated stirred tank reactor ($D_{\mathrm{STR}} = 180\,\mathrm{mm}$ or $k \approx 35\,\mathrm{m}^{-1}$) and it is limited in the upper range by the local resolution of 0.0015 m/voxel ($k \approx 4188\,\mathrm{m}^{-1}$). This also explains the strong drop at $k \approx 3000\,\mathrm{m}^{-1}$, because with the current resolution of the data no smaller structures (<1.5 mm) can be resolved and thus the energy for those structures is underrepresented in the energy spectrum. Therefore, the drop cannot be attributed to the viscous region, since this is present only from the KOLMOGOROV length of about $k > 6500\,\mathrm{m}^{-1}$ (at an averaged energy dissipation rate of $\varepsilon_{\mathrm{T}} = 0.012\,\mathrm{W/Kg}$ for this configuration at $n = 250\,\mathrm{rpm}$). The local peak by $k \approx 2100$ can be explained by the fact that due to the spatially limited resolution the energy of the smallest structures cannot be assigned to the respective scales. The energy of the smaller scales is shifted towards the larger scales and leads to an apparent local peak in the absolute energy.

Alternatively, the local peak in Figure 4.9 at a wavenumber of $k \approx 2100\,\mathrm{m}^{-1}$ can be explained by the tailing vortices introduced by the Rushton turbine. The wavenumber of $k \approx 2100\,\mathrm{m}^{-1}$ corresponds to half the blade height of a Rushton stirrer ($\approx 3\,\mathrm{mm}$), which is illustrated in Figure 4.10. Based on the work of Linek et al. [Lin75], Figure 4.10 schemati-

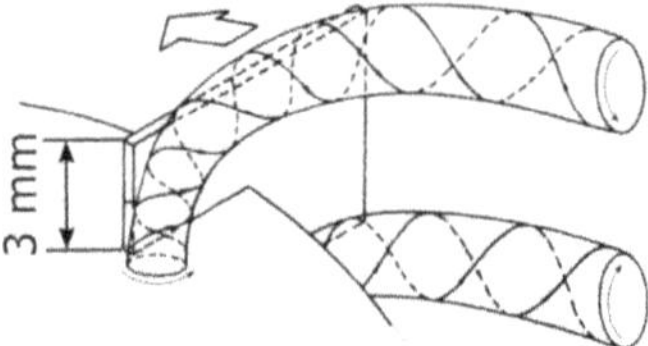

Figure 4.10. Schematic visualization of the trailing vortex behind a Rushton turbine blade [Lin75]

cally shows the trailing vortex behind a blade of a Rushton stirrer, which is created on the upper and lower side of the Rushton turbine. In addition, it was demonstrated in Kuschel at al. [Kus21] that the highest flow velocities are generated in these vortices by superposition of the rotation of the stirrer blade and the vortex. It has been shown that not only the scale of the stirrer diameter but also the smaller scales (k_2 and k_3) have a recognizable influence on the energy spectrum. This explains the deviation of the energy cascades from the theoretical slope of $-5/3$: the theory assumes that energy is transferred into the system on a single scale rather than multiple scales.

Table 4.1 lists the integral values for the specific power inputs based on the torque measurements and the 4DPTV measurements. For 4DPTV measurements, the absolute kinetic energy in each subvolume was calculated based on the instantaneous velocity fields and then averaged in time and space.

Table 4.1. Comparison of the determined specific power input P based on the torque M and flow field measurements

Stirrer frequency n /rpm	**specific power input $P/V/\mathrm{W}\cdot\mathrm{m}^{-3}$ based on:**	
	torque measurement	**4DPTV measurement**
250	12.6	11.2
350	34.6	41.1
450	73.4	59.2

The specific power inputs determined with the 4DPTV method do not exactly match those of the torque method. However, the comparison based on the specific power input shows that the measured Lagrangian trajectories and the calculated instantaneous velocity fields are plausible and can be used for further analysis.

4.1.2. Characterization of Stirred Tank Reactors by Means of the Local Mixing Time Distribution

In the following, the results on the local mixing time distribution in a laboratory-scale 3 L stirred tank reactor are presented. For this purpose, the influence of observation perspective on the local mixing time distribution is first discussed. Subsequently, the influence of the stirrer frequency as well as the aeration on the mixing in the stirred tank reactor is visualized and discussed.

Influence of the Observation Direction on the Local Mixing Time Measurements

Since the determination of the local mixing time distribution is based on the recording of a color change experiment using the backlight illumination method, it must first be checked whether there is a dependence on the perspective. For this purpose, same mixing time experiments are evaluated from two different angles for demonstration purposes. For these experiments, two identical cameras, lenses and LED panels are used, each with the same operating settings (see Chapter 3.2).

Figure 4.11 shows an example of the relative local mixing time distribution for an experiment from two perspectives A and B. The considered planes are orthogonal to each other. In Figure 4.11, the location of the addition of the diluted hydrochloric acid solution is visualized by a red dot. In both perspectives (A & B) in Figure 4.11, the development of the macroscopic mixing can be clearly observed starting from the place of addition. Moreover, in both visualizations in Figure 4.11 a high similarity of the local mixing time distribution can be observed. However, these are subjective observations. Therefore in Figure 4.12, the respective relative local mixing time distribution is plotted as a histogram. This visualization provides a much more objective comparison. Based on the very high similarity of the two density distributions for perspective A and B and the same global mixing time, it can be reasonably assumed that the mixing time is not expected to depend on the angle under consideration. Furthermore, due to the pH indicator used, light from the LED panels is selectively absorbed by the pH tracer depending on the pH value. Therefore it can be assumed that the two perspectives are independent of each other.
The relative probability frequency distributions shown in Figure 4.12 have a bimodal distribution for both perspectives. This is an indication that the mixing processes in the stirred tank reactor can be divided into two compartments. The first compartment is formed by the volume above the upper stirrer and the second compartment is formed by the volume

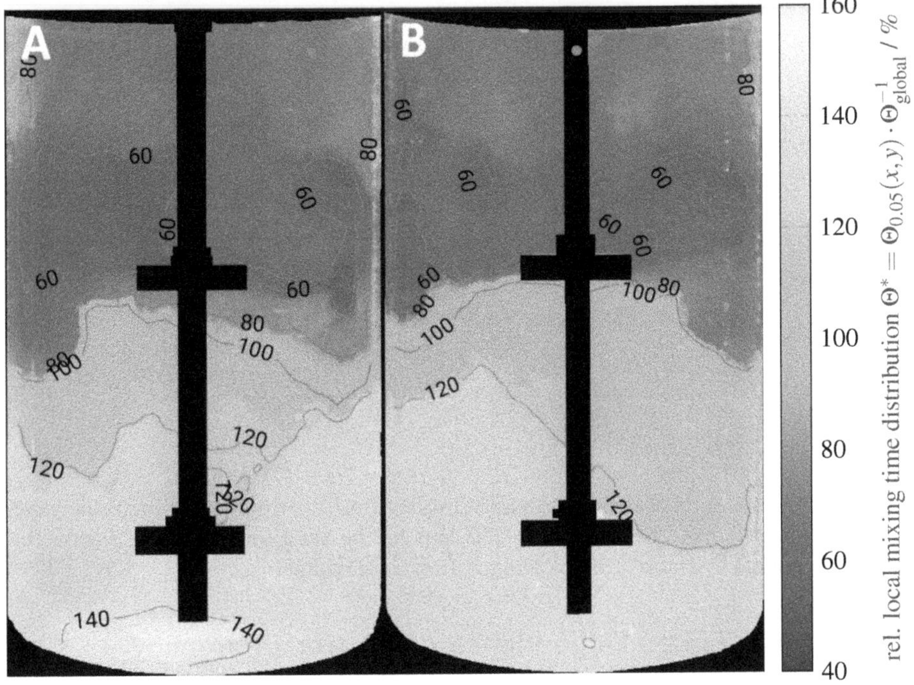

Figure 4.11. Comparison of the projected local mixing time distribution for the same experiment ($n = 250$ rpm and no aeration) in the 3 L reactor for a double Rushton turbine configuration from two perspectives (A & B); red dot: point of addition

below the lower stirrer. The volume between the two stirrers shows a transition between both compartments. To quantify these compartments, the mean value of that range (0 - 100 and 100 - 200 %) is represented by a red dashed line in Figure 4.12. Considering the results of the flow field measurement, the two areas of bimodal distribution are due to two coherent compartments in the system caused by the stirrer. The mean mixing time of the first compartment in the example shown here is $t_{\text{mix},1} \approx 7.8\,s$ and for the second compartment $t_{\text{mix},1} \approx 14.6\,s$.

Based on the comparison between the two perspectives shown in Figure 4.12, no significant deviations in the determined relative local mixing time distribution can be detected. Based on the visual comparison in Figure 4.11 and the objective comparison in Figures 4.11 & 4.12, it can be concluded that the mixing in the stirred tank reactor is nearly independent of the perspective. For this reason, only one perspective is further examined in the following.

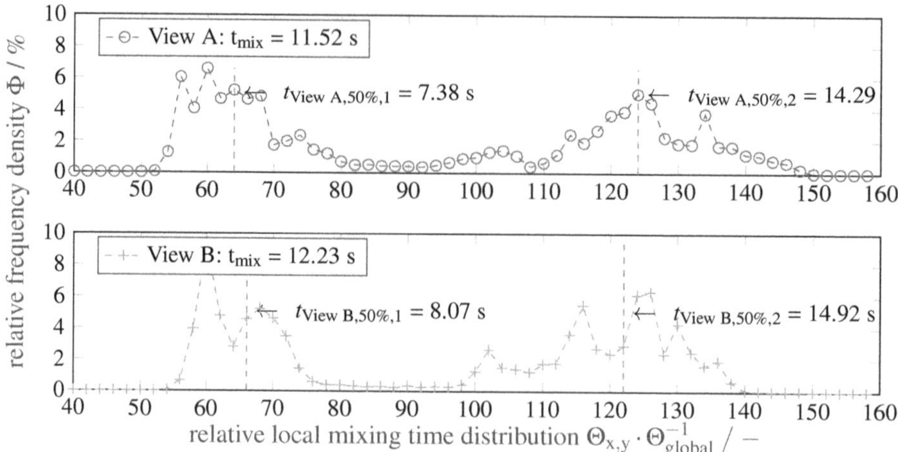

Figure 4.12. Comparison of the relative local mixing time distribution for the same experiment (stirrer frequency $n = 250$ rpm and no aeration) in the 3 L reactor for a double Rushton turbine configuration from two perspectives (A & B)

Local Mixing Time: Dependence on Stirrer Frequency - unaerated

First, the single-phase results are presented and discussed. Subsequently, the two-phase results are considered. The single-phase mixing time of a system is of particular importance and must always be taken into account when designing a system. The mixing time provides information about how quickly the local concentrations in the system equalize, for example after the addition of a reactant or a nutrient. Furthermore, if the system is designed correctly, the global mixing time is a quality parameter that guarantees consistent production conditions by ensuring uniform and sufficient homogenization.

Figure 4.13 shows the relative local mixing time distribution Θ^* maps for the double Rushton turbine configuration and four different stirrer frequencies. In addition, the relative frequency distribution of the absolute mixing time is shown in Figure 4.14 and the relative frequency distribution of the relative local mixing time Θ^* is shown in Figure 4.15. On the basis of Figure 4.13, a large similarity between the relative local mixing time Θ^* distribution maps and thus a small dependence on the stirrer frequency can be seen. This becomes especially clear based on the relative frequency Φ distribution in Figure 4.15. In Figure 4.14, the relative frequency distribution Φ is plotted against the absolute local mixing time Θ.

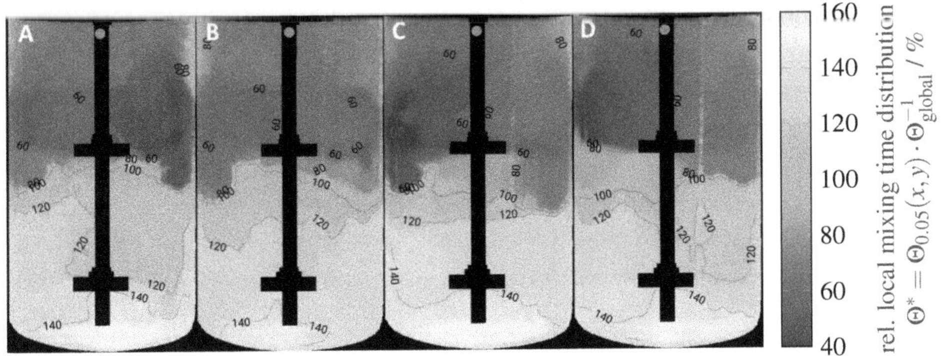

Figure 4.13. Comparison of the local mixing time distribution for different stirrer frequencies: A) $n = 150$ rpm, B) $n = 250$ rpm, C) $n = 350$ rpm, D) $n = 450$ rpm; 3L STR; 2x Rushton turbine; red dot: point of addition

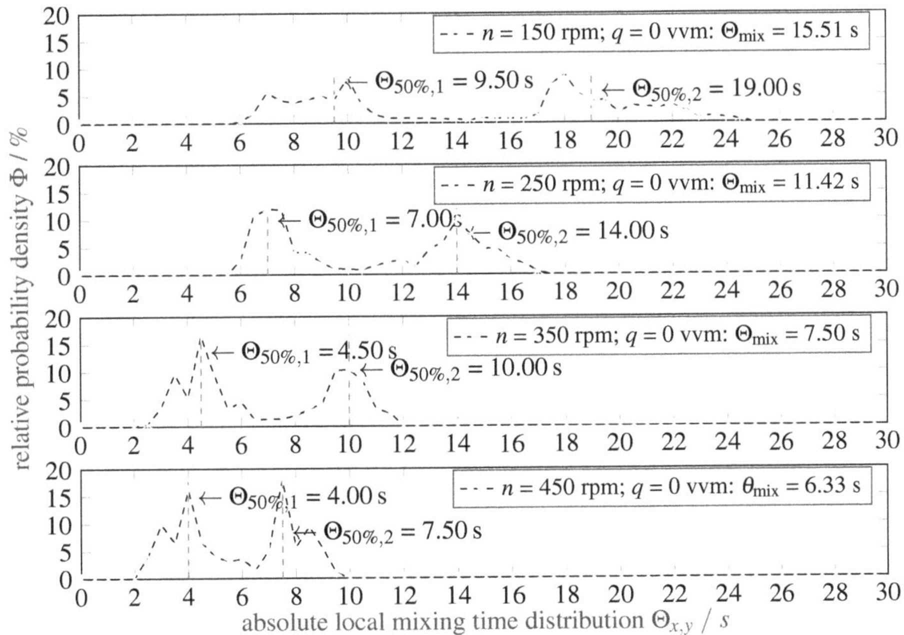

Figure 4.14. Comparison of the absolute local mixing time distribution for different stirrer frequencies ($n = [150, 250, 350, 450]$ rpm and no aeration) 3L STR; double Rushton turbine

Figure 4.14 indicates that with increasing stirrer frequency the global mixing time decreases, and the local mixing time distribution is shifted to smaller values. Furthermore, the relative frequency distribution of the absolute local mixing time appears to become more narrow with increasing stirrer frequency. However, if the local mixing time will be normalized by the corresponding global mixing time the relative frequency density distribution appears to be equal for all four stirrer frequencies. Therefore, it can be concluded that the global single phase mixing time scales with the stirrer frequency, although the prevailing compartments (based on the relative mixing time maps) are independent from the stirrer frequency. This

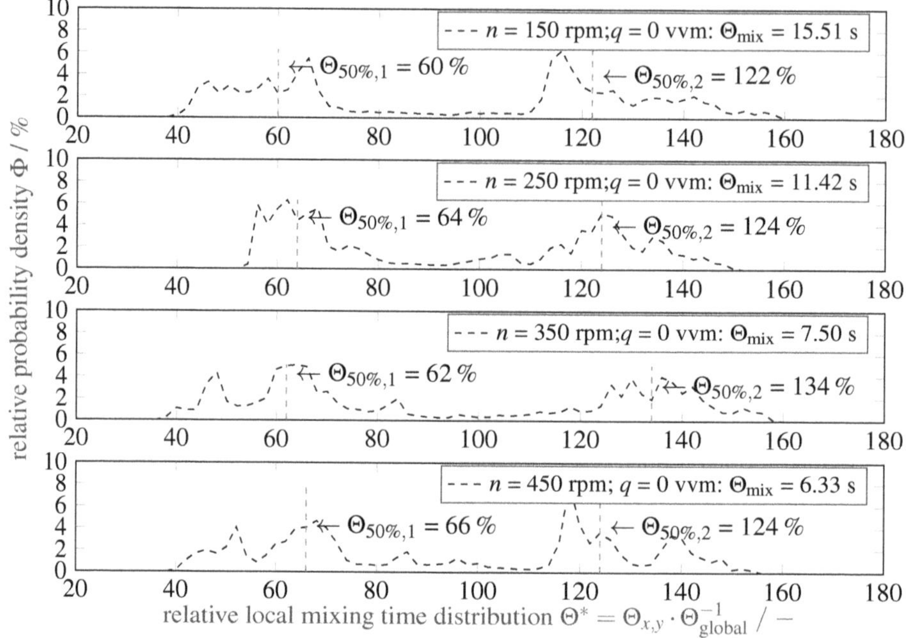

Figure 4.15. Comparison of the relative local mixing time distribution for different stirrer frequencies ($n = [150, 250, 350, 450]$ rpm and no aeration) in the 3L reactor; class width: 1 %p

leads to the conclusion that the method of the relative mixing time distribution is a pure characterization of the geometrical parameters of the system, since obviously no direct dependence on the stirrer frequency can be seen. The relative deviation in mixing time of the respective compartments can be directly calculated from the global mixing time. In the example shown, the average relative local mixing time equals $\overline{\Theta^*} \approx 63\,\%$ for the first compartment and $\overline{\Theta^*} \approx 125\,\%$ for the second compartment. However, this relationship can only be used to a limited extent to describe multi-stage stirrer systems in general, since the described relationship depends not only on the geometric parameters but also strongly on the feeding position. Nevertheless, the local mixing time distribution map can be used to represent and quantify the coherent structures that are mainly influenced by the mixing of the system. A transfer of the method to larger systems would be of particular interest to gain a better understanding of the reactors in terms of scale transfer. However, full optical and a non-distorted access to systems with larger reactor volumes becomes much more difficult, since an appropriate water basin for refractive index matching cannot easily be built around the reactor.

Local Mixing Time: Dependence on Stirrer Frequency - aerated

Due to the fact that most industrially used stirred tank reactors are operated as gas/liquid systems, an extension of the local mixing time determination to such processes is presented in the following. In principle, the described method for the single-phase case can be used for this purpose.
Figure 4.16 initially shows the global mixing times for the investigated operating conditions with aeration compared to the unaerated operation. It becomes obvious that the influence

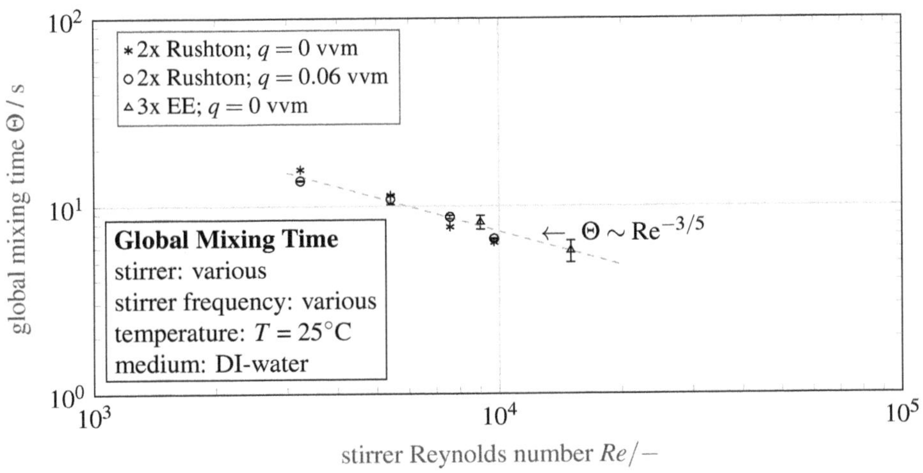

Figure 4.16. Comparison of the global mixing time Θ_{global} for different stirrer configurations, without and with aeration in a 3 L STR

of aeration on the global mixing time is very small. Only for the lowest stirrer frequency can an influence be determined based on a shorter mixing time compared to the unaerated condition. Otherwise, it can be assumed that the mechanical power input is dominant over the buoyancy driven flows, because only the specific power input from the gas phase is comparable to the specific power input through the stirrer for the lowest stirrer frequency. For a volumetric aeration rate of $q = 0.6\,\text{vvm}$, the specific power input of the gas phase is $P/V_{\text{buoyancydriven}} = \rho \cdot g\Delta h = 1.9\,\text{W}\,\text{m}^{-3}$ [Web21]. Whereas the specific power input of the double Rushton stirrer at a stirrer frequency of $n = 150\,\text{rpm}$ is already $P/V = 2.5\,\text{W}\,\text{m}^{-3}$ and increases to $P/V = 68.2\,\text{W}\,\text{m}^{-3}$ at $n = 450\,\text{rpm}$. Furthermore, it is evident that the mixing time decay scales very closely with $\sim Re^{-3/5}$ for both stirrer configurations.

Figure 4.17 shows the local mixing time distribution map for the single-phase case (see Figure 4.17 first row) as well as for the two-phase operation with a gassing rate of $q = 0.06\,\text{vvm}$ (see Figure 4.17 second row) for four different stirrer frequencies. By comparing

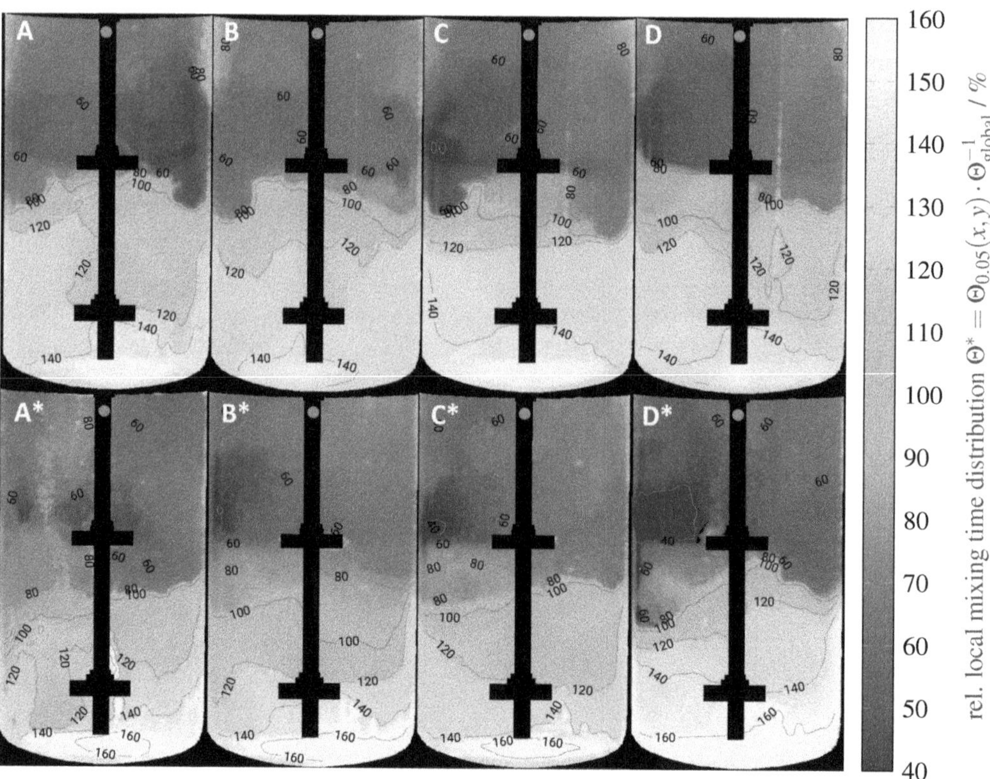

Figure 4.17. Comparison of the relative local mixing time Θ^* maps for different stirrer frequencies A) $n = 150$ rpm, B) $n = 250$ rpm, C) $n = 350$ rpm, D) $n = 450$ rpm; without (A - D) and with (A* - D*) aeration in a 3 L STR; red dot: point of addition

the aerated (A*, B*, C*, D*) and unaerated (A, B, C, D) operating conditions based on the local mixing time results in Figure 4.17, the effect of aeration on the coherent structures in the system can be clearly seen. This is particularly evident when comparing the operating states A and A* at the lowest stirrer frequency $n = 250$ rpm. It can be seen that the contour lines (red) are shifted and the surfaces of one color are broken up. It can be shown that the previously shown coherent structures can be broken up by the gas phase and that the mixing time is positively influenced by the inhomogeneities in the system. A representation of the gas phase distribution is shown in Figure 4.18 and illustrates the influence of the gas phase on the mixing structures. Figure 4.18 clearly shows that impeller flooding occurs at an stirrer frequency of $n = 150$ rpm. The resulting bouncy driven flow leads to a large-scale flow structure. From a stirrer frequency of $n \geq 350$ rpm, the gas phase is homogeneously distributed and impeller loading occurs without gas recirculation.

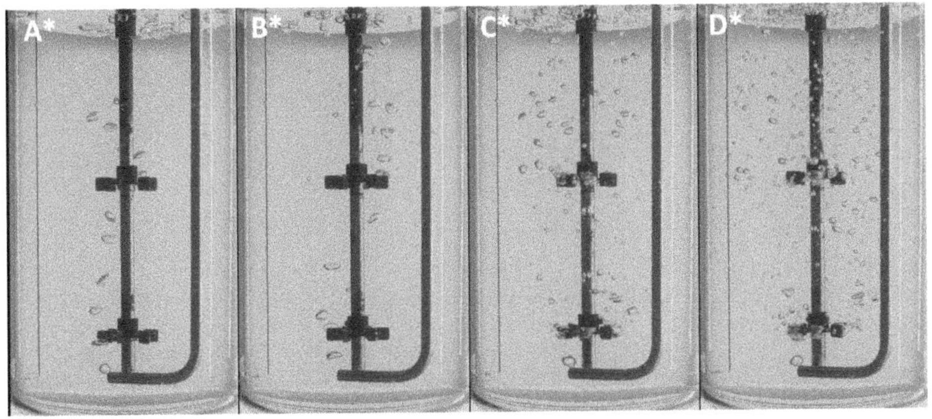

Figure 4.18. Schematic visualization of the gas phase distribution for different stirrer frequencies A*) $n = 150$ rpm, B*) $n = 250$ rpm, C*) $n = 350$ rpm, D*) $n = 450$ rpm; with aeration in a 3 L STR

Furthermore, the effect of the gas phase on mixing can also be observed in the reduction of the global mixing time ($\Delta\Theta_{global} \approx 2$s in Figure 4.16). The observed reduction of the global mixing time is in fact not statistically significant, however, it is an indication that in the 3 L stirred tank reactor, the effect described by Rosseburg et al. [Ros18], that a heterogeneous flow reduces the mixing time, also applies. Furthermore, it can be seen that at the highest stirrer frequency $n = 450$ rpm the local mixing time distribution maps for both operating conditions have converged again. This is due to the fact that the specific power is dominant over the buoyancy driven flows.

Furthermore, based on the representation in Figure 4.17, the mixing time history can be described impressively well and better or worse mixed zones can be identified as well. The closer the lines of constant mixing time (red) are next to each other, the higher the mixing time gradient between these regions. Conversely, this means that these regions represent a higher mass exchange barrier than other regions in the system.

4.1.3. Characterization of Stirred Tank Reactors by Means of the Local Propagation Time

As already discussed theoretically in the chapter 2.1.3, the measurement of the mixing time can be used not only to determine the local mixing times but also to resolve the local illumination gradients. To resolve the local illumination gradients the mixing time is not used directly, but the notion of local spreading of the tracer will be introduced. For this purpose, analogous to the determination of the residence time distribution in continuously operated stirred tank reactors, the gradient of the temporal change in the gray value is used. The cumulative measured value thus corresponds to the average time period required until a change at the input of the system has an effect on the measuring point.

According to control theory, the behavior of a system can be very well determined by either the impulse or step response method. In the following, it will be shown that the experimental procedure for the local macro-mixing time method is very similar to the step response method and thus considerations from control engineering can be adapted. Likewise, the experiments for mixing time measurement are used for the following evaluation. Due to the double excess of added hydrochloric acid for the mixing time measurement, the used pH indicator assumes either the blue (alkaline) or yellow (acidic) state (see Figure 2.13). Therefore, it can be assumed that the pH indicator behaves like a binary (0 or 1) system. Thus, the addition of the acid leads to a jump from the color blue to yellow and thus corresponds to a step response measurement. For this purpose, the distribution function $\Psi(t)$

$$\Psi(n,m,t) = \frac{\mathrm{d}}{\mathrm{d}t} \frac{I(n,m,t)}{I(n,m,t \to \infty)} \tag{4.1}$$

can be defined with the time course of the gray value $I(t)$ and the final value $I(t \to \infty)$. By multiplying the distribution function $\Psi(t)$ by the sampling time t, the distribution characteristic propagation time

$$\tau(x,y) = \int_0^\infty t\Psi(n,m,t)\mathrm{d}t \tag{4.2}$$

is obtained for each pixel x and y in the image. Based on the assumptions made and the shown effect that the considered problem is rationally symmetric (see chapter 4.1.2), the distribution of the mean propagation time τ as well as the spatial gradient θ of the propagation time is shown as a distribution map in Figure 4.19 and 4.20 and respectively. Figure 4.19 shows the local mean propagation times τ for the unaerated operating condition (first row) and the aerated operating condition (second row) for four different stirrer frequencies. By means of the characteristic propagation time maps shown in Figure 4.19, it can be seen that

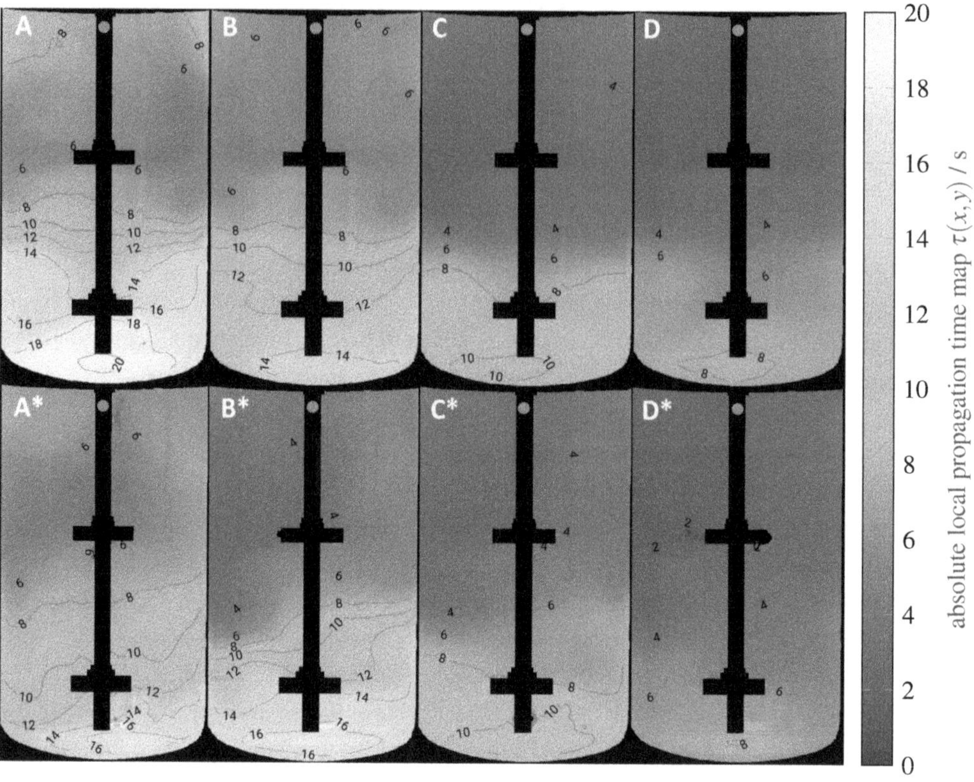

Figure 4.19. Comparison of the absolute local distribution time for different stirrer frequencies A) $n = 150$ rpm, B) $n = 250$ rpm, C) $n = 350$ rpm, D) $n = 450$ rpm; without (A - D) and with (A* - D*) aeration in a 3 L STR; red dot: point of addition

propagation starts from the addition point and spreads uniformly from the addition point (1 cm below the water surface) downwards in the unaerated operating state. It is particularly noticeable that the isolines in the area between the two agitators are very close together. This clearly shows the reduced mass exchange between the two characteristic vortices of the two stirring elements. Whereas the longest propagation times always occur under the lowest stirrer. Furthermore, the selected representation of the characteristic propagation time clearly shows that the coherent structures can be broken up by the aeration and that the maximum propagation time decreases significantly from 20 to 16 s.

Figure 4.20 shows the spatial gradient of the propagation times $d\tau \cdot dx$ for the unaerated operating condition (first row) and the aerated operating condition (second row) for four different stirrer frequencies. The spatial gradient is a direct measure of how fast the added

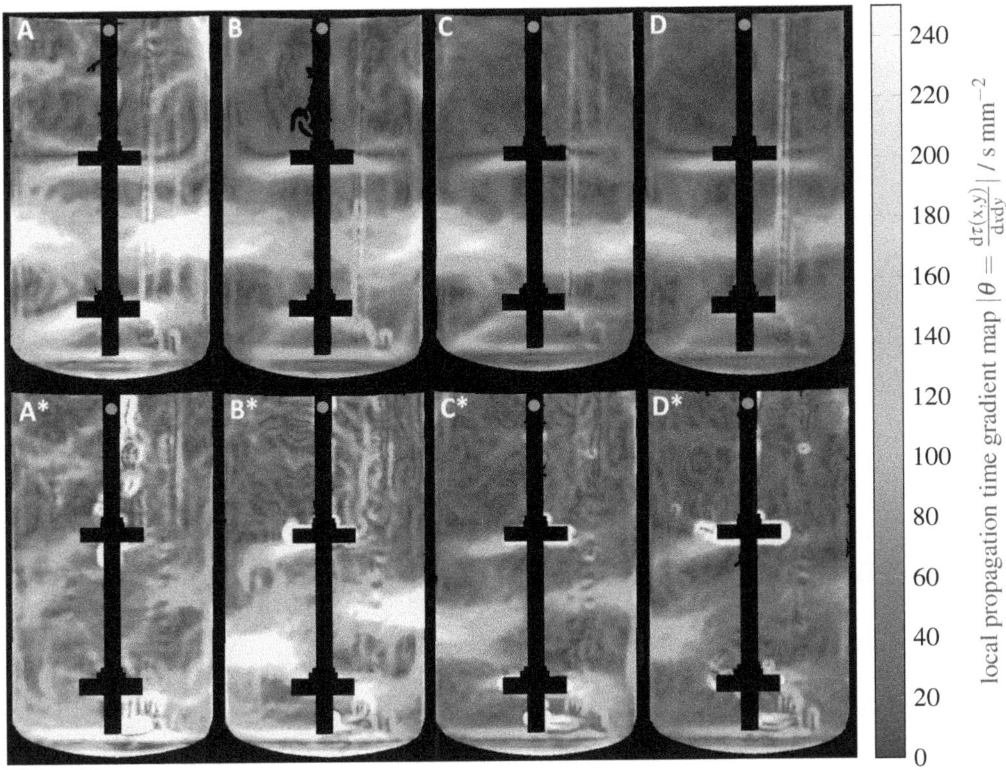

Figure 4.20. Comparison of the absolute local mixing time distribution for different stirrer frequencies A) $n = 150$ rpm, B) $n = 250$ rpm, C) $n = 350$ rpm, D) $n = 450$ rpm; without (A - D) and with (A* - D*) aeration in a 3 L STR; red dot: point of addition

acid spreads in the system. The greater the value, the slower the spreading (Reciprocal of the velocity; unit s/mm). This can be used to identify poorly mixed zones (like in the upper right corner in Figure 4.20 A), as well as to identify barriers between coherent structures (between the two stirrers in Figure 4.20 A-D). Furthermore, it is clear from Figure 4.20 that the local gradients can be either shifted (A* & B*) or even be reduced (C* & D*) by aeration.

4.2. Lagrangian Analysis of Particle Trajectories

In the following, the Lagrangian analyses are presented. For this purpose, the measured particle trajectories are evaluated first. Then the velocity fields and Sojourn times are calculated. Finally, dispersion coefficients for stirred tank reactors are shown and discussed.

Figure 4.21 shows an example of a selection of particle trajectories. The color code of the trajectories refers to the modified velocity magnitude

$$< |\boldsymbol{v}_{\mathrm{mod.}}| >=< |\boldsymbol{v}| > \cdot f_{\mathrm{d}} \tag{4.3}$$

with the velocity magnitude $|\boldsymbol{v}|$ and the binary vertical direction

$$f_{\mathrm{d}} =< \boldsymbol{Y}(t=t_{\mathrm{i+1}}) - \boldsymbol{Y}(t=t_{\mathrm{i}}) >= \begin{cases} 1, & \text{if } f_{\mathrm{d}} > 0 \\ -1, & \text{if } f_{\mathrm{d}} < 0 \end{cases} \tag{4.4}$$

which is calculated for each location in the space based on the vertical position $\boldsymbol{Y}$ of each trajectories. Here, a negative velocity means that the particle moves downwards (blue) and

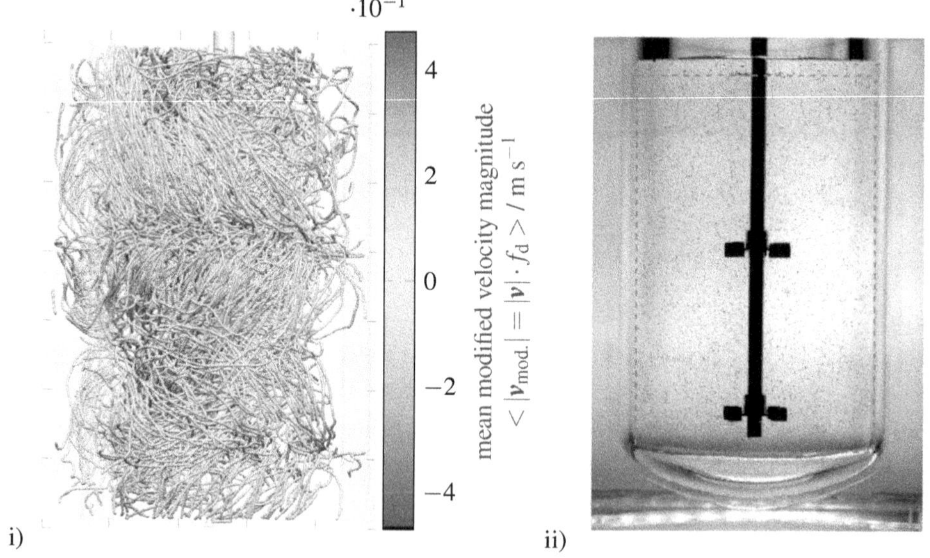

Figure 4.21. Representation of 1% of all trajectories for the double Rushton turbine configuration at a stirrer frequency of $n = 250$ rpm; a negative modified velocity magnitude indicates that the trajectory moves downwards

a positive velocity means that it moves upwards (red). Figure 4.22 shows the number of particle trajectories with a certain temporal track length for the experiments with the double Rushton turbine i) and triple Elephant Ear stirrer ii) configurations, respectively. In this case,

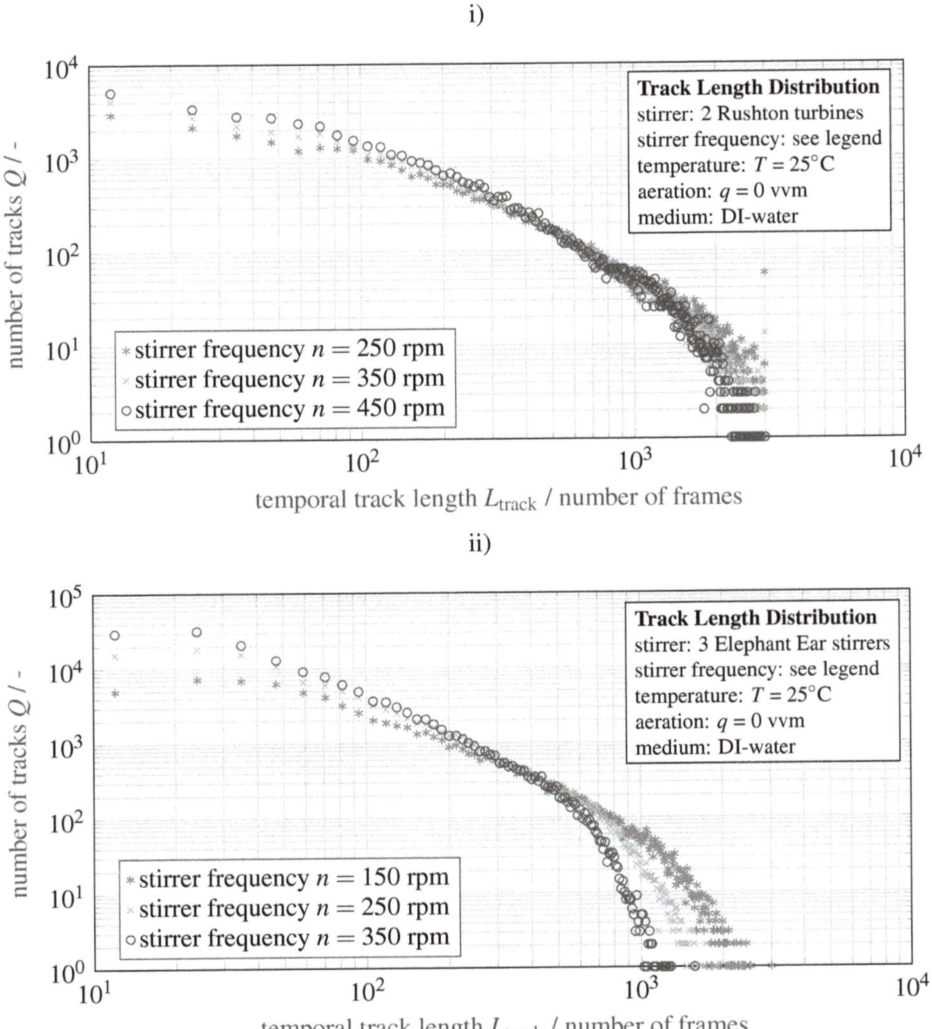

Figure 4.22. Comparison of the absolute track length for the double Rushton stirrer i) and triple Elephant Ear ii) configuration at different frequencies and no aeration) in the 3 L reactor; class width: 8 bit resolution between 0 - 3000

the indicated length refers to the time period of the observation. The longer the track length

the better the measurement and the following statistical evaluations.
In Figure 4.22 i) and ii) a minimum length of 6 time steps (depending on the iterations in the Shake-The-Box evaluation) and 3000 (maximum number of frames) is possible. Especially for the experiments with the double Rushton turbine, it can be seen that many trajectories could be observed over the total measurement time. Due to the significantly larger Elephant Ear stirrer and the resulting shadowing, the particle trajectories are shorter on average. Neverthelesss, the particle trajectories are on average between 260 - 300 time steps long, which corresponds to a time span Δt of 0.33 - 0.38 s or $\approx 10\,\%$ of the total measurement duration.
Figure 4.23 i) illustrates the temporally and azimuthally averaged velocities directly from the

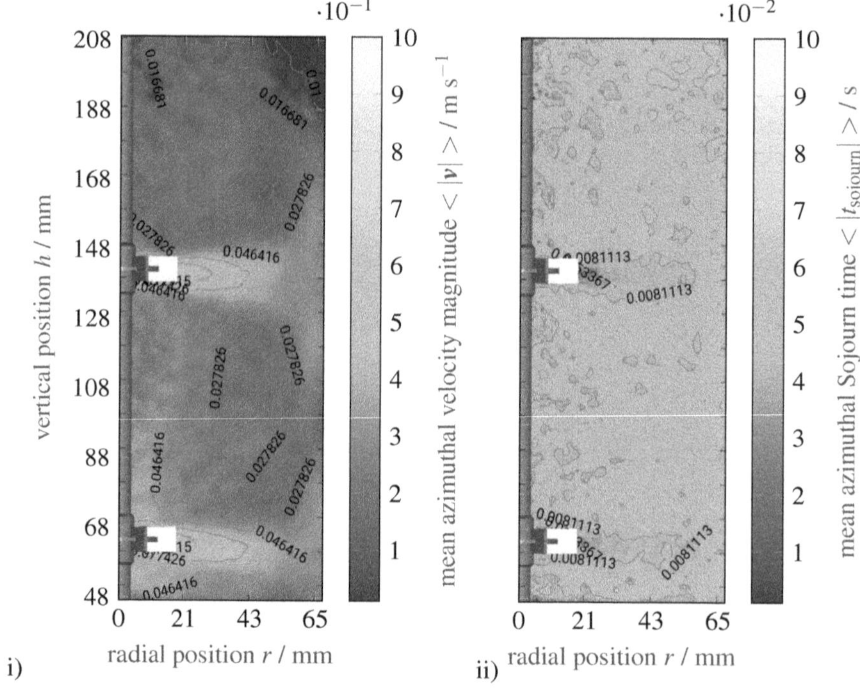

Figure 4.23. i) Time and azimuthally averaged velocity map and ii) time and azimuthally averaged Sojourn time t_{sojourn}; double Rushton configuration; $n = 250$ rpm; no aeration

Lagrangian data. For this purpose, the Lagrangian data have been projected directly onto the Eulerian grid and then time-averaged. Furthermore, in Figure 4.23 ii) the local mean Sojourn time t_{sojourn} is shown. In Figure 4.23, the Lagrangian data are projected onto a 1x1x1 mm^3 subvolume and then averaged azimuthally between two baffles.
In principle, the Sojourn time t_{sojourn} corresponds to the reciprocal of the velocity with respect

to the subvolume size. However, this only applies if a particle passes through the subvolume without deviating from the main flow direction. As soon as the particle is on a curved path or trapped in a small vortex structure, the local residence time increases. Moreover, when projecting Lagrangian trajectories onto the Eulerian grid, the problem arises that the residence time is significantly underestimated if the trajectory only intersects the subvolume, because this seems to reduce the characteristic length of the subvolume.

Based on the Lagrangian trajectories, the mean exposure time period for which the particle collective experiences a certain acceleration and thus external force, can be given directly. This information is of particular interest for processes that are sensitive to external forces. In Figure 4.24, the calculated mean exposure time spans are plotted against the absolute acceleration with an 8 bit resolution for the x-axis values. Figure 4.24 shows that at a constant

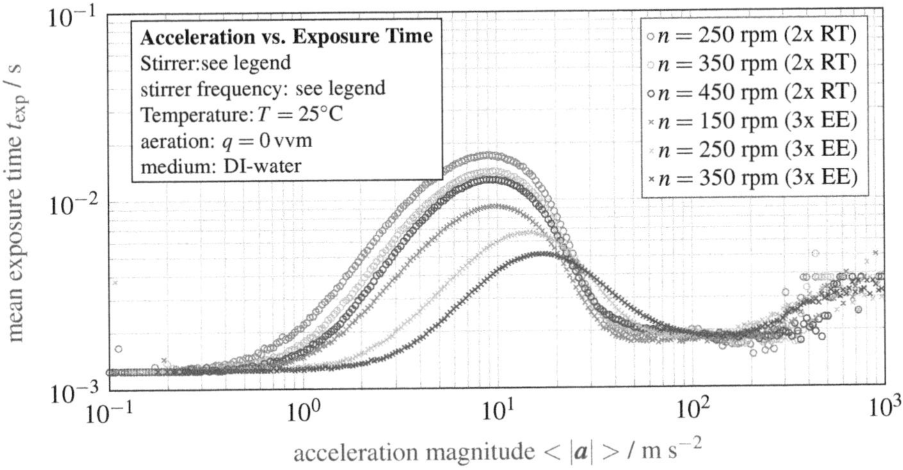

Figure 4.24. Comparison of the mean exposure time t_{exp} as a function of the acceleration magnitude $< |\boldsymbol{a}| >$ for the double Rushton turbine (RT) and the Elephant Ear stirrer (EE) configuration at different frequencies in the 3 L reactor; class width: 8 bit resolution between 0 - 3000

acceleration magnitude $< |\boldsymbol{a}| >$, the mean exposure time t_{exp} decreases with increasing stirrer frequency. This means that with a constant mean exposure time, the respective acceleration values increase. This causal relationship is plausible, since the turbulence in the system is intensified with increasing stirrer frequency. Furthermore, it can be seen that with increasing stirrer frequency, the acceleration values become larger, which a particle experiences for the longest time on average statistically.

4.2.1. Determination of Absolute Particle Dispersion in Stirred Tank Reactors

The results of the single particle analysis are presented below. In order to ensure sufficiently good statistics, those particle trajectories which are too short are excluded from the evaluation. However, the definition of "too short" is not unambiguous and must be checked separately for each measurement. For this purpose, the Lagrange autocorrelation function $F(t)$ is calculated in an iterative process, and the time period considered is continuously increased. As soon as the autocorrelation function tends to zero, the iteration is stopped and the characteristic Lagrange integration time T_{L} is calculated. Subsequently, only those particle trajectories are considered for the following analyses, which are longer than the Lagrange integration time T_{L}. The autocorrelation function is shown in Figure 4.25 for the stirrer frequency $n = 250$ rpm and four different time spans as an example. Based on the track

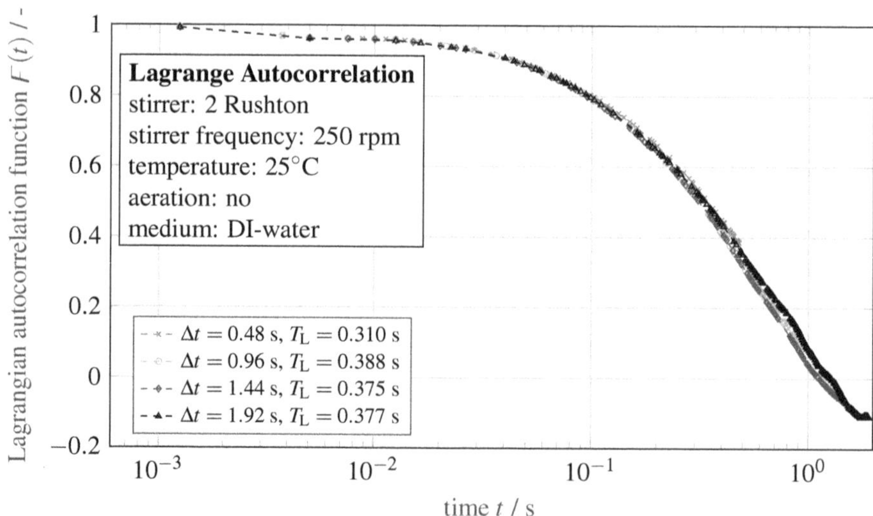

Figure 4.25. Visualization of the Lagrange autocorrelation function $F(t)$ for the double Rushton turbine configuration and a stirrer frequency of $n = 250$ rpm for different evaluation time frames Δt

lengths shown in Figure 4.22 the problem arises that the number of trajectories which can be evaluated over the evaluation time span decreases drastically. If the evaluation time span chosen is too short, the equilibrium shifts drastically in the direction of the significantly more frequently represented shorter trajectories. For this reason, an optimum must be determined between a maximum observation period and a minimum number of particle trajectories.

Based on the data shown in Figure 4.22, an observation time span of $\Delta t = 0.96$ s seems to be sufficient. For the following evaluations a time span of $\Delta t(n) = 0.96 \cdot 250/(60 \cdot n)$ is chosen based on the results shown in Figure 4.22.

Using the example of the double Rushton turbine configuration at a strirrer frequency $n = 250$ rpm, the mean particle displacement is shown in Figure 4.26. The mean particle

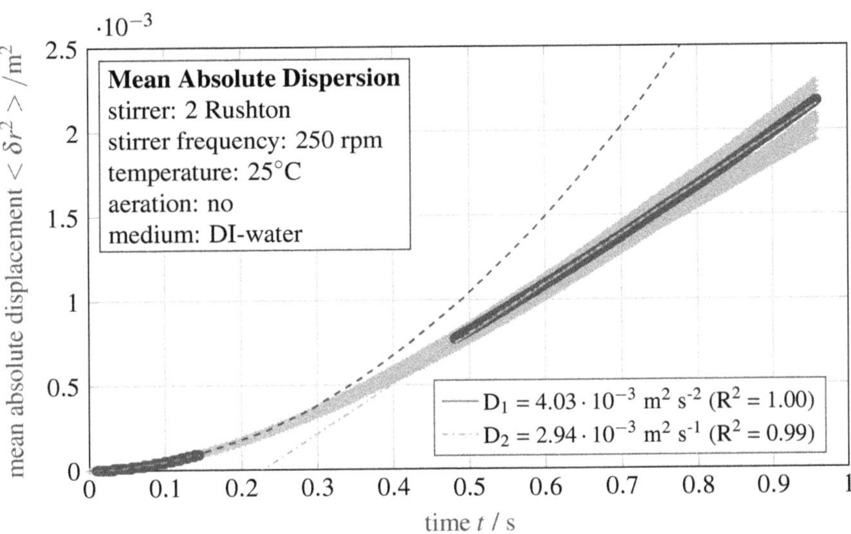

Figure 4.26. Visualization of the mean absolute particle displacement $< \delta r^2 >$ vs. sampling time t for the double Rushton turbine configuration and a stirrer frequency of $n = 250$ rpm; $T_L = 0.388$s

displacement is a measure of how far each particle moves on average from its point of origin. In addition, the time spans for the evaluation of the ballistic and turbulent diffusive dispersion coefficients D_1 and D_2 (according to equations 2.22 & 2.23) are marked in blue in Figure 4.26.

For a better overview, the results for the determination of the respective dispersion coefficients for all stirrer frequencies and geometries are summarized in Figure 4.27. It can be seen that the ballistic D_1 and turbulent diffusive dispersion D_2 coefficients for the Elephant Ear stirrer system are significantly higher than the values for the system with two Rushton stirrers. This is due to the significantly larger stirring elements and the better mixing resulting in the system. Due to the fact that the different stirring elements have different diameters, the stirrer tip speeds u_{tip} for the two systems and the investigated frequencies are not the same. For this reason, the dimensionless dispersion coefficients are shown in the following Figure 4.28.

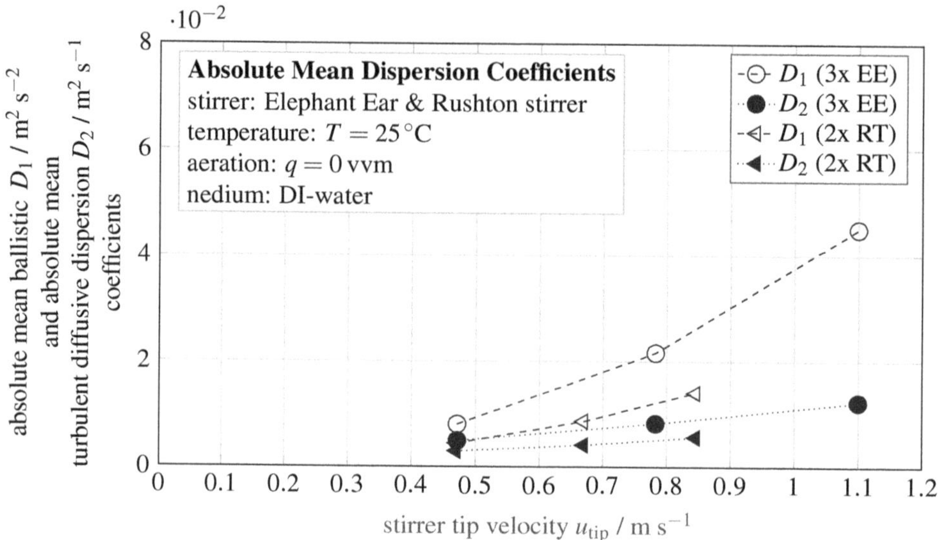

Figure 4.27. Mean ballistic D_1 and turbulent diffusive D_2 dispersion coefficients for the investigated stirrer geometries as a function of stirrer tip speed u_{tip}

For this purpose, the dispersion coefficients are related to the stirrer tip speed u_{tip} and the stirrer diameter d_{stirrer}. Based on Figure 4.28 it is evident that the dimensionless dispersion coefficients are constant in the range of the investigated stirrer frequencies. Moreover, it can be observed especially that the dimensionless turbulent diffusive dispersion coefficient seems to be almost the same for both stirrer configurations. For this reason, an average proportionality factor can be given for both configurations, which allows the prediction of the turbulent diffusive dispersion coefficient for different stirrer frequencies n and both stirrer configurations.

$$D_2(n) = \overline{D_2^*}(u_{\text{tip}}) \cdot u_{\text{tip}} \cdot d = 0.177 \pm 0.006 \cdot \pi \cdot n \cdot d^2 \tag{4.5}$$

Based on Equation 4.5, it is now possible to determine the turbulent diffusive dispersion coefficients directly from the dimensionless system-independent dispersion coefficient. In particular, the turbulent diffusive dispersion coefficient is important for the characterization of the system, since it is a direct measure of the large time scale mass transport in the system. In general, the larger the turbulent diffusive dispersion, the more intensive the mixing in the system.

Furthermore, the dispersion coefficients can also be spatially calculated and illustrated. Figure 4.29 illustrates the ballistic (i)) and the turbulent diffusive (ii)) dispersion coefficients

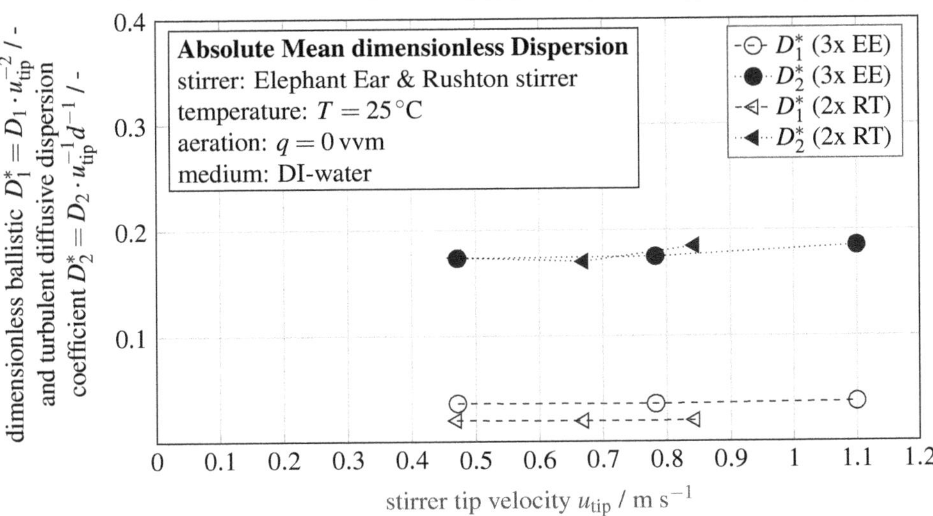

Figure 4.28. Mean dimensionless ballistic D_1^* and turbulent diffusive D_2^* dispersion coefficients for the investigated stirrer geometries as a function of stirrer tip speed u_{tip}

locally resolved. For the spatial visualization, the dispersion coefficients are calculated for each particle trajectory and then the values are assigned to the respective starting points of the trajectory. For better representation, the spatial data are then azimuthally averaged. The evaluated volume corresponds to the section of the reactor shown in Figure 4.1. For a better understanding of the shown spatial dispersion coefficients, on the one hand the dimensionless dispersion coefficients are shown (according to Figure 4.28) and on the other hand it is additionally considered for the representation whether the end point of the particle trajectories are above or below the staring point. For this, a vertical binary direction factor

$$f_d = < \boldsymbol{Y}(t = t_{end}) - \boldsymbol{Y}(t = t_0) > = \begin{cases} 1, & \text{if } f_d > 0 \\ -1, & \text{if } f_d < 0 \end{cases} \tag{4.6}$$

is calculated for each location in the space based on the vertical start $\boldsymbol{Y}(t = t_0)$ and end $\boldsymbol{Y}(t = t_{end})$ position of each trajectory, where red means an upward ($f_d = 1$) dispersion and blue means a downward ($f_d = -1$) dispersion. Due to shading around the baffles and stirrers, the trajectories in these areas are not long enough, so dispersion coefficients cannot be calculated at these locations (black). The spatial dispersion coefficients were evaluated on an

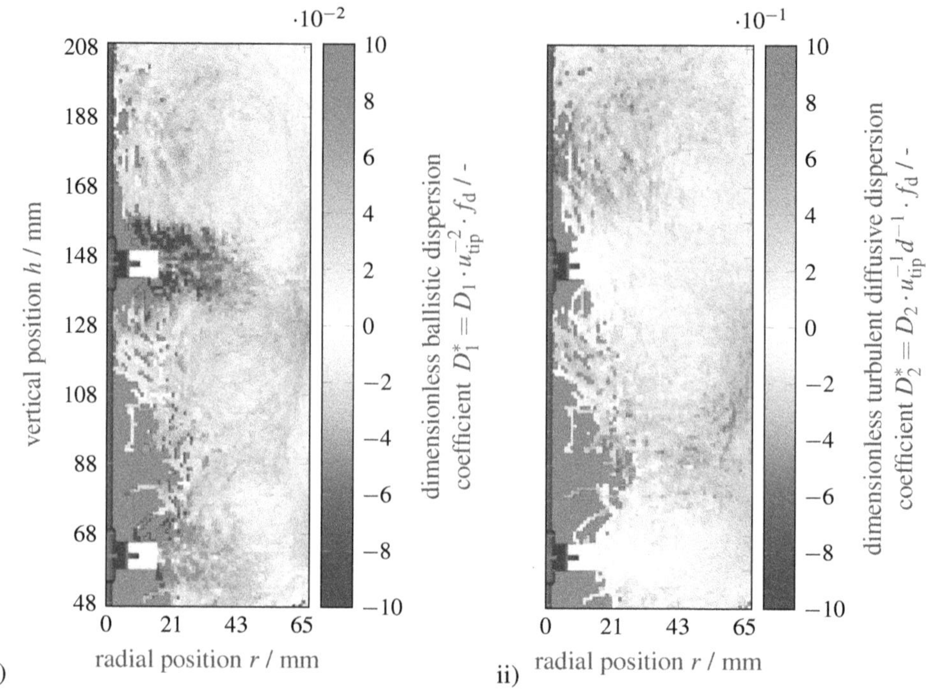

Figure 4.29. Modified spatial ballistic i) and turbulent diffusive ii) dispersion coefficients (red: particle moves upwards, blue: particle moves downwards, gray: no information); double Rushton configuration; $n = 250\,\text{rpm}$; no aeration; observation time $\Delta t \approx 1\,\text{s}$; number of observed trajectories $N_{\text{tracks}} \approx 1100$

average of 10 particles per spatial location ($\Delta V = 1 \cdot 1 \cdot 1\,\text{mm}^3$) represented.

Due to the fact that the ballistic dispersion coefficient is determined for very short time spans ($t \ll T_{\text{L}}$), the magnitude of the ballistic dispersion coefficient must correlate with the flow velocity. This can be seen in Figure 4.29 i), the largest values for the ballistic dispersion coefficient are located in the downstream flow of the two Rushton stirrers. The fact that both red and blue points appear in the area above and below the upper Rushton turbine is due to the fact that the trajectories are first accelerated radially outward from the Rushton turbine and then deflected upward (red) or downward (blue) at the inner wall of the reactor. The fact that the red/blue distribution is the same in both figures (see Figure 4.29 i) ii)) is due to the fact that the evaluation of the two dispersion coefficients is performed along a respective trajectory, and thus the vertical direction is the same for both dispersion coefficients. Only the intensity of the blue and red points differs due to the different dispersion coefficients.

Figure 4.29 ii) shows the spatial turbulent diffusive dispersion coefficients. It can be seen that, in contrast to the ballistic dispersion coefficients, the values around the Rushton stirrers are among the lowest. This is an indication that the trajectories within the observed period of $\Delta t \approx 1$ s come back close to the respective starting point. Furthermore, this is a clear indication of temporally stable coherent structures in the system. Furthermore, on the vertical position of $h \approx 90$ mm, a clear horizontal boundary between red and blue starting points can be seen as well as high dispersion coefficients. This means that the trajectories from this location show a very clear tendency back upwards or downwards, respectively. Additionally, a comparison of the local dimensionless turbulent diffusive dispersion for three different stirrer frequencies is shown in Figure 4.30. The comparison impressively shows how similar

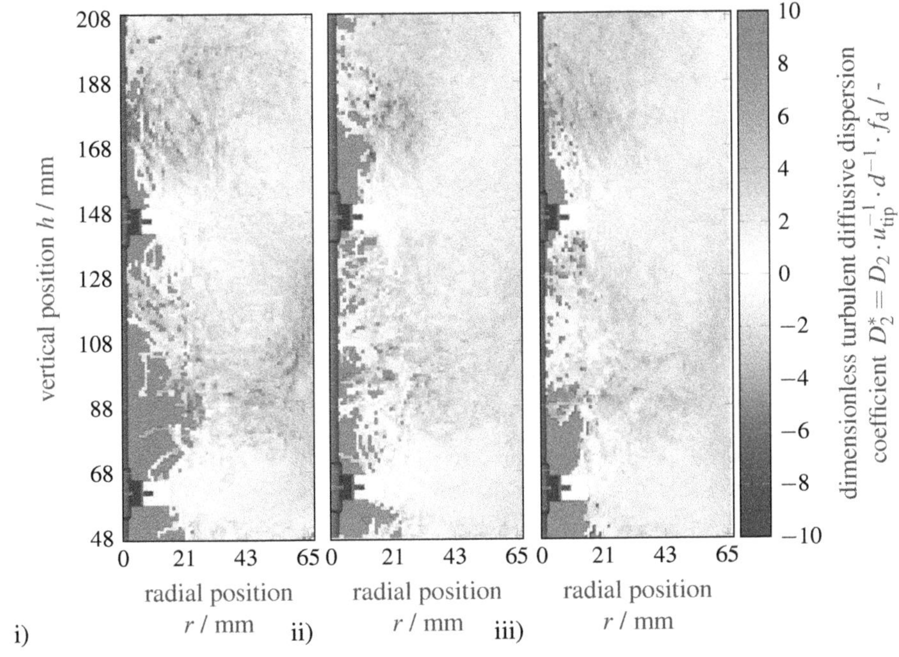

Figure 4.30. Modified spatial turbulent diffusive dispersion coefficient; (red: particle moves upward, blue: particle moves downwards, black: no information); double Rushton configuration; i) $n = 250$ rpm, ii) $n = 350$ rpm, iii) $n = 450$ rpm; no aeration; observation time $\Delta t \approx 1$ s; number of observed trajectories $N_{tracks} \approx 1100$

the structures are in the investigated stirred tank reactor, because the distribution of local dimensionless turbulent dispersion shown in Figure 4.29 ii) appears to be very similar for all three stirring frequencies. This is very strong evidence of coherent structures in the stirred tank reactor.

In Figure 4.31 iii) the helical mixing based on the dimensionless turbulent diffusive dispersion distribution is drawn as a dashed line as an example and is compared with the corresponding flow map in Figure 4.31 ii). In comparison with the flow fields for this configuration, it can be seen that the two characteristic vortices of the two Rushton stirrers intersect at a vertical position of $h \approx 90\,\mathrm{mm}$. The separation of the two compartments at this point can now also be confirmed qualitatively based on the turbulent diffusive dispersion.

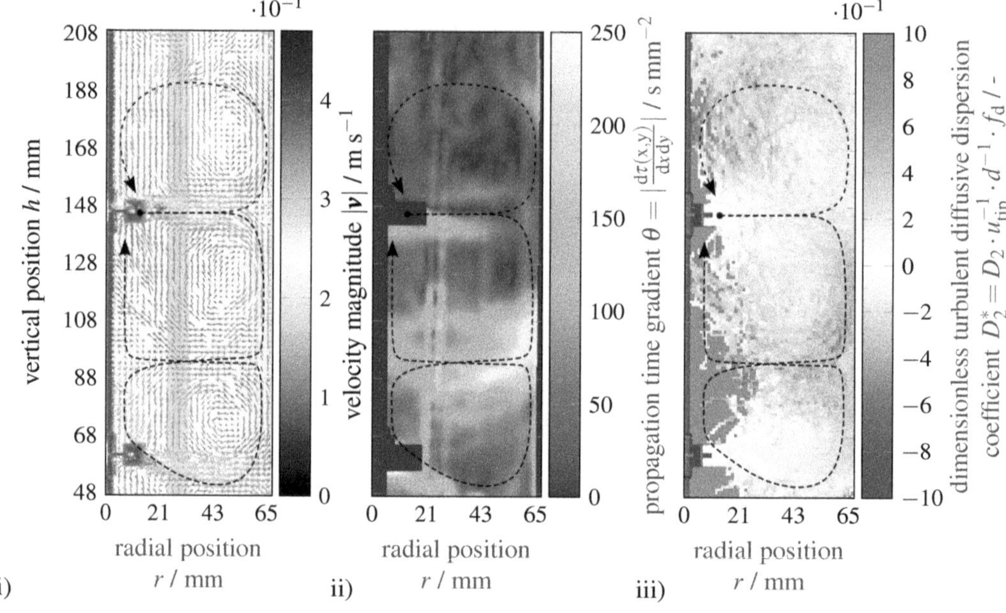

Figure 4.31. Comparison of the schematic mixing structure based on the spatial dimensionless turbulent diffusive dispersion with the corresponding time-averaged flow field

Another view on the temporally coherent structures based on the turbulent diffusive dispersion coefficient is given in Figure 4.32 as a three-dimensional representation. For the three-dimensional representation, the local dispersion coefficients are shown as iso-volumes, with the values around $D_2^* = 0.177 \pm 0.006$ shown transparently to provide a better overview. The coherent structures which contribute to the compartment formation become impressively clear again on the basis of the three-dimensional representation. In addition, it also becomes obvious what significant influence baffles have on mixing. Especially in Figure 4.32 i), it can

Figure 4.32. Visualization of the three dimensional distribution of the dimensionless turbulent diffusive dispersion coefficient; i) front view, ii) side view I, iii) back view, iv) side view II; direction of rotation: mathematically negative seen from top view

be seen that the highest dispersion values are in front of the baffles. This can be explained by the fact that the tangential flow (tangential to the direction of rotation of the stirrers) is deflected upwards/downwards at the baffles.

In summary, it can be said that the consideration of absolute dispersion in relation to the characterization of transport processes in stirred tank reactors is still completely new. However, this method has the potential to provide much more detailed insight into the transport processes of stirred tank reactors. Furthermore, the evaluation of the spatial dispersion coefficients has two great advantages. First, the local turbulent diffusive dispersion coefficients can be used to quantify the flow fields with respect to mixing. Secondly, coherent structures can be identified which have a negative influence on the global mixing processes in the system, and should therefore be avoided.

5. Conclusion

Process equipment characterization is an important tool which improves the understanding of existing equipment and enables transfer to new equipment and processes. This work has intensively studied the possibilities of hydrodynamic characterization of an aerated stirred tank reactor using the example of mixing time distribution and Lagrangian particle dispersion to identify heterogeneities.

This thesis is motivated by the question of how Lagrangian analysis approaches can contribute to a better understanding of aerated stirred tank reactors with respect to mass transport and mixing. For this purpose, common methods for the determination of integral parameters for equipment characterization, such as the determination of the specific power input and the global mixing time, were described and discussed. Furthermore, the method of local mixing time determination was used to describe the investigated stirrer configurations with respect to the mixing structures and to provide a comparative data basis for the following Lagrangian analyses. For the Lagrangian analyses, the Lagrangian particle tracking approach was chosen. Using the 4DPTV method, high temporal and spatial resolution velocity fields as well as the three-dimensional trajectories of the tracer particles have been resolved.

Initially, the two stirrer geometries used in this thesis were investigated in detail with regard to their power input characteristics using the torque method and compared with the literature. A good agreement has been found for the determined power input coefficients.

For further hydrodynamic characterization to quantify heterogeneities, the local mixing time distribution was determined first. It could be shown that in single-phase operation there is a very large self-similarity of the mixing structures at different power inputs. From this it could be concluded that coherent structures are present in the system. With the extended evaluation of the mixing time experiments with regard to the local propagation time of the introduced tracer, it was also possible to show how these structures can be broken up depending on the mechanical power input in two-phase operation. The observed improvement in the propagation time was a total of 4 s or 20% compared to the single-phase operating condition.

Furthermore, it could be clearly shown which regions in the stirred tank reactor tend to represent a barrier to mass transport and which regions have good mixing behavior. The local gradient of the propagation time of the tracer was used for this purpose. In particular, the zones between the two radial stirring elements were identified as strong barriers. Nonetheless, the consideration of the local mixing and propagation time had been based on the Eulerian approach, so that only stationary information about coherent structures could be determined and quantified. For this reason, the Lagrangian approach to hydrodynamic characterization was also applied in this work. For this purpose, spatially and temporally highly resolved 4DPTV measurements were performed and evaluated using the very efficient and reliable "shake the box" method.

Due to the fact that the 4DPTV method is relatively new compared to other methods for flow field measurements and has never been used before to determine Lagrangian trajectories and Eulerian flow fields in the entire stirred tank reactor volume, the results are first validated using the absolute kinetic energy. For this purpose, the absolute kinetic energy based on the three-dimensional transient flow fields was compared with the torque measurements. Although a smaller deviation between the two power number determination methods could be observed, it was found that the transient three-dimensional flow fields and thus the 4DPTV method was successfully implemented and the evaluated data were valid as well. This means not only that the transient three-dimensional flow fields are valid, but also that the particles used follow the flow ideally, so that further analyses of the trajectories are possible.

First, the trajectories were analyzed with respect to the local residence time in the system and compared with the time-averaged flow fields. Due to the fact that the residence time distribution must in principle correspond to the reciprocal of the local velocity distribution, a good agreement in the residence time distribution structures could be identified, assuming that further analyses are valid. Based on the residence time distribution, the respective exposure of the particles to a certain external acceleration force was determined. The comparison of the stirrer geometries on the basis of the exposure time distribution showed that the highest external forces increase with increasing specific power input, but they were significantly greater with the Elephant Ear stirrer than with the Rushton stirrer configuration in the 3 L stirred tank reactor. This is a very interesting result, since Elephant Ear stirrers were previously considered to be particularly less shear intensive. It should be noted that, it is not possible to infer the shear rate in the flow directly from the external acceleration force on a particle. It is, however, contrary to expectations that the external forces and the exposure of these forces to a particle were much lower for the Rushton stirrer configuration, although the specific power input was greater than in the Elephant Ear stirrers.

Furthermore, the absolute dispersion was determined on the basis of the Lagrangian trajectories. First, the validity of the Lagrangian autocorrelation function was checked in order to determine the Lagrangian autocorrelation time. Afterwards, the ballistic and turbulent diffusive dispersion coefficients were determined for both stirrer configurations. Based on the dispersion coefficients, it could be shown that the dispersion coefficients become larger with increasing power input and the values for the Elephant Ear configuration are larger than for the Rushton configuration. However, it could also be shown that the turbulent diffusive dispersion coefficients for both systems correlate linearly with the stirrer tip velocity. On the contrary, it has even been shown that the turbulent diffusive dispersion coefficients could be successfully nondimensioned and a constant parameter $\overline{D_2^*}(u_{\mathrm{tip}}) = 0.177 \pm 0.006$ can be specified for both systems. In this way, it is possible to determine the system specific and power input dependent turbulent diffusive dispersion coefficients independently for both systems. Furthermore, with the three-dimensional representation of the turbulent diffusive dispersion coefficients, the mixing structures and barriers could be visualized, which contributes to a better understanding of aerated stirred tank reactors.

The experimental results shown in this work have been successfully validated and provide significantly deeper insights into heterogeneous structures of aerated stirred tank reactors. Furthermore, Lagrangian analysis approaches have been successfully applied as a complement to established characterization methods. Based on the results and methods presented in this thesis, not only can existing processes be investigated with respect to heterogeneous structures, a more efficient reactor design can also be achieved. Furthermore, the Lagrangian approach allows the analysis of real particle trajectories and thus the three-dimensional mass transport. Nonetheless, the application of Lagrangian analysis for the hydrodynamic characterization of stirred tank reactors is still in the development phase, and therefore its application is currently limited to laboratory scale reactors. This gap and the gap in our ability to adapt processes to larger scales will need to be closed in the coming years.

Bibliography

[Ala08] Alam, J.M. and Lin, J.C. *Toward a Fully Lagrangian Atmospheric Modeling System. Monthly Weather Review*, 136(12):4653 – 4667, 2008. doi:10.1175/2008MWR2515.1.

[Ale18] Alexakis, A. and Biferale, L. *Cascades and transitions in turbulent flows. Physics Reports*, 767-769:1–101, 2018. ISSN 0370-1573. doi:https://doi.org/10.1016/j.physrep.2018.08.001. Cascades and transitions in turbulent flows.

[Alv02] Alvarez, M.M., Arratia, P.E. and Muzzio, F.J. *Laminar mixing in eccentric stirred tank systems. Can. J. Chem. Eng.*, 80:546, 2002.

[Bak08] Bakunin, O.G. *Turbulence and Diffusion*, vol. 1. Springer, Berlin, Heidelberg, 2008. doi:https://doi.org/10.1007/978-3-540-68222-6.

[Bou83] Bourne, J. *Mixing on the Molecular Scale (Micromixing). Chemical Engineering Science*, 38(1):5–8, 1983. ISSN 00092509. doi:10.1016/0009-2509(83)80129-8.

[Cab07] Cabaret, F., Bonnot, S., Fradette, L. and Tanguy, P.A. *Mixing Time Analysis Using Colorimetric Methods and Image Processing. Industrial & Engineering Chemistry Research*, 46(14):5032–5042, 2007. doi:10.1021/ie0613265.

[Cam03] Campolo, M., Sbrizzai, F. and Soldati, A. *Time-dependent flow structures and Lagrangian mixing in Rushton-impeller baffled-tank reactor. Chemical Engineering Science*, 58(8):1615–1629, 2003. ISSN 0009-2509. doi:https://doi.org/10.1016/S0009-2509(02)00658-9.

[DS16] D. Schanz, S. Gesemann, A.S. *Shake-The-Box: Lagrangian particle tracking at high particle image densities. Experiments in Fluids volume*, 2016. doi:https://doi.org/10.1007/s00348-016-2157-1.

[Ell12] Ellis, G. *Chapter 9 - Filters in Control Systems.* In G. Ellis, editor, *Control System Design Guide (Fourth Edition)*, pp. 165–183. Butterworth-Heinemann, Boston, fourth edition edn., 2012. ISBN 978-0-12-385920-4. doi:https://doi.org/10.1016/B978-0-12-385920-4.00009-6.

[Enr19] Enrile, F., Besio, G., Stocchino, A. and Magaldi, M.G. *Influence of initial conditions on absolute and relative dispersion in semi-enclosed basins. PLOS ONE*, 14(7):1–12, 2019. doi:10.1371/journal.pone.0217073.

[Fit21] Fitschen, J., Hofmann, S., Wutz, J., Kameke, A., Hoffmann, M., Wucherpfennig, T. and Schlüter, M. *Novel evaluation method to determine the local mixing time distribution in stirred tank reactors. Chemical Engineering Science: X*, 10:100098, 2021. ISSN 2590-1400. doi:https://doi.org/10.1016/j.cesx.2021.100098.

[Fre10] Freund, R.J., Wilson, W.J. and Mohr, D.L. *CHAPTER 1 - Data and Statistics.* In R.J. Freund, W.J. Wilson and D.L. Mohr, editors, *Statistical Methods (Third Edition)*, pp. 1–65. Academic Press, Boston, third edition edn., 2010. ISBN 978-0-12-374970-3. doi:https://doi.org/10.1016/B978-0-12-374970-3.00001-9.

[Gab09] Gabriele, A., Nienow, A. and Simmons, M. *Use of angle resolved PIV to estimate local specific energy dissipation rates for up- and down-pumping pitched blade agitators in a stirred tank. Chemical Engineering Science*, 64(1):126–143, 2009. ISSN 0009-2509. doi:https://doi.org/10.1016/j.ces.2008.09.018.

[Gez00] Gezork, K., Bujalski, W., Cooke, M. and Nienow, A. *The Transition from Homogeneous to Heterogeneous Flow in a Gassed, Stirred Vessel. Chemical Engineering Research and Design*, 78(3):363 – 370, 2000. ISSN 0263-8762. doi: http://dx.doi.org/10.1205/026387600527482. Fluid Mixing.

[Gmb21] GmbH, E.H. *Ekato Stirrer*, 2021.

[GO04] Garcia-Ochoa, F. and Gomez, E. *Theoretical prediction of gas–liquid mass transfer coefficient, specific area and hold-up in sparged stirred tanks. Chemical Engineering Science*, 59(12):2489 – 2501, 2004. ISSN 0009-2509. doi: http://dx.doi.org/10.1016/j.ces.2004.02.009.

[God62] Godleski, E.S. and Smith, J.C. *Power requirements and blend times in the agitation of pseudoplastic fluids. AIChE J.*, 8:617, 1962.

[Hal15] Haller, G. *Lagrangian Coherent Structures. Annual Review of Fluid Mechanics*, 47(1):137–162, 2015. doi:10.1146/annurev-fluid-010313-141322.

[Her06] Herwig. *Grundgleichungen der Strömungsmechanik*, pp. 47–80. Springer Berlin Heidelberg, Berlin, Heidelberg, 2006. ISBN 978-3-540-32443-0. doi:10.1007/3-540-32443-7_4.

[Hib79] Hiby, J.W. *Definition und Messung der Mischgüte in flüssigen Gemischen. Chemie Ingenieur Technik*, 51(7):704–709, 1979. doi:https://doi.org/10.1002/cite.330510705.

[Hud89] Hudcova, V., Machon, V. and Nienow, A.W. *Gas–liquid dispersion with dual Rushton impellers. Biotechnology and Bioengineering*, 34(5):617–628, 1989. doi: https://doi.org/10.1002/bit.260340506.

[Huh12] Huhn, F., von Kameke, A., Pérez-Muñuzuri, V., Olascoaga, M.J. and Beron-Vera, F.J. *The impact of advective transport by the South Indian Ocean Countercurrent on the Madagascar plankton bloom. Geophysical Research Letters*, 39(6), 2012. doi:https://doi.org/10.1029/2012GL051246.

[HX13] H. Xia, N. Francois, H.P.M.S. *Lagrangian scale of particle dispersion in turbulence. Nature Communications*, 2013. doi:https://doi.org/10.1038/ncomms3013.

[Kam19] Kameke, A., Kastens, S., Rüttinger, S., Herres-Pawlis, S. and Schlüter, M. *How coherent structures dominate the residence time in a bubble wake: An experimental example. Chemical Engineering Science*, 207:317–326, 2019. ISSN 0009-2509. doi:https://doi.org/10.1016/j.ces.2019.06.033.

[Kra64] Kraichnan, R.H. *Kolmogorov's Hypotheses and Eulerian Turbulence Theory. The Physics of Fluids*, 7(11):1723–1734, 1964. doi:10.1063/1.2746572.

[Kra02] Kraume, M. *Homogenisieren in Rührbehältern*, chap. 2, pp. 21–44. John Wiley Sons, Ltd, 2002. ISBN 9783527603367. doi:https://doi.org/10.1002/3527603360.ch2.

[Kra12] Kraume, M. *Mischen und Rühren*, pp. 555–601. Springer Berlin Heidelberg, Berlin, Heidelberg, 2012. ISBN 978-3-642-25149-8. doi:10.1007/978-3-642-25149-8_18.

[Kra14] Kraume, M. *Die Entwicklung der Rührtechnik von einer empirischen Kunst zur Wissenschaft. Chemie Ingenieur Technik*, 86(12):2051–2062, 2014. doi:https://doi.org/10.1002/cite.201400124.

[Kre02] Krebs, R. *Bauelemente rührtechnischer Apparate — Auslegungskriterien, Wirtschaftlichkeit, anwendungsorientierte Lösungen*, chap. 7, pp. 147–174. John Wiley Sons, Ltd, 2002. ISBN 9783527603367. doi:https://doi.org/10.1002/3527603360.ch7.

[Kus20] Kuschel, M. and Takors, R. *Simulated oxygen and glucose gradients as a prerequisite for predicting industrial scale performance a priori. Biotechnology and Bioengineering*, 117(9):2760–2770, 2020. doi:https://doi.org/10.1002/bit.27457.

[Kus21] Kuschel, M., Fitschen, J., Hoffmann, M., von Kameke, A., Schlüter, M. and Wucherpfennig, T. *Validation of Novel Lattice Boltzmann Large Eddy Simulations (LB LES) for Equipment Characterization in Biopharma. Processes*, 9(6), 2021. ISSN 2227-9717. doi:10.3390/pr9060950.

[Lin75] Linek, V. and Smith, J.M. *The trailing vortex system produced by Rushton turbine agitators. Chemical Engineering and Science*, 30:1093–1105, 1975.

[Lla20] Llamas, C.G., Spille, C., Kastens, S., Paz, D.G., Schlüter, M. and von Kameke, A. *Potential of Lagrangian Analysis Methods in the Study of Chemical Reactors. Chemie Ingenieur Technik*, 92(5):540–553, 2020. doi:https://doi.org/10.1002/cite.201900147.

[Man94] Manikowski, M., Bodemeier, S., Lübbert, A., Bujalski, W. and Nienow, A.W. *Measurement of Gas and Liquid Flows in Stirred Tank Reactors with Multiple Agitators. The Canadian Journal of Chemical Engineering*, 72(5):769–781, 1994. ISSN 00084034, 1939019X. doi:10.1002/cjce.5450720502.

[Mav01] Mavros, P. *Flow Visualization in Stirred Vessels. Chemical Engineering Research and Design*, 79(2):113 – 127, 2001. ISSN 0263-8762. doi:http://dx.doi.org/10.1205/02638760151095926.

[Mew02] Mewes, D. *Optische und tomographische Messverfahren für Mischprozesse*, chap. 4, pp. 63–89. John Wiley Sons, Ltd, 2002. ISBN 9783527603367. doi:https://doi.org/10.1002/3527603360.ch4.

[Mic16] Michaelis, D., Neal, D.R. and Wieneke, B. *Peak-locking reduction for particle image velocimetry. Measurement Science and Technology*, 27(10):104005, 2016. doi:10.1088/0957-0233/27/10/104005.

[Mon15] Montante, G. and Paglianti, A. *Gas hold-up distribution and mixing time in gas–liquid stirred tanks. Chemical Engineering Journal*, 279:648 – 658, 2015. ISSN 1385-8947. doi:http://dx.doi.org/10.1016/j.cej.2015.05.058.

[Mor73] Morel, P. and Bandeen, W. *the Eole experiment: early results and current objectives. Bulletin of the American Meteorological Society*, 54(4):298 – 306, 1973. doi:10.1175/1520-0477-54.4.298.

[Nau15] Nauha, E.K., Visuri, O., Vermasvuori, R. and Alopaeus, V. *A new simple approach for the scale-up of aerated stirred tanks. Chemical Engineering Research and Design*, 95:150 – 161, 2015. ISSN 0263-8762. doi:http://dx.doi.org/10.1016/j.cherd.2014.10.015.

[Nor07] *DIN 28 130: Chemical equipment - Survey on components of agitator vessels with agitation unit*, 2007.

[Ram01] Rammohan, A., Kemoun, A., Al-Dahhan, M. and Dudukovic, M. *A Lagrangian description of flows in stirred tanks via computer-automated radioactive particle tracking (CARPT). Chemical Engineering Science*, 56(8):2629–2639, 2001. ISSN 0009-2509. doi:https://doi.org/10.1016/S0009-2509(00)00537-6.

[Ric22] Richardson, L.F. *Weather prediction by numerical process.* London: Cambridge Univ. Press, XII u. 236 S. 4° (1922)., 1922.

[Rod13] Rodriguez, G., Weheliye, W., Anderlei, T., Micheletti, M., Yianneskis, M. and Ducci, A. *Mixing time and kinetic energy measurements in a shaken cylindrical bioreactor. Chemical Engineering Research and Design*, 91(11):2084 – 2097, 2013. ISSN 0263-8762. doi:https://doi.org/10.1016/j.cherd.2013.03.005. Mixing.

[Ros18] Rosseburg, A., Fitschen, J., Wutz, J., Wucherpfennig, T. and Schlüter, M. *Hydrodynamic inhomogeneities in large scale stirred tanks - Influence on mixing time. Chemical Engineering Science*, 188:208 – 220, 2018. ISSN 0009-2509. doi:https://doi.org/10.1016/j.ces.2018.05.008.

[Sie11] Sieblist, C., Jenzsch, M., Pohlscheidt, M. and Lübbert, A. *Insights into Large-Scale Cell-Culture Reactors: I. Liquid Mixing and Oxygen Supply. Biotechnology Journal*, 6(12):1532–1546, 2011. ISSN 18606768. doi:10.1002/biot.201000408.

[Som10] Sommer. *Micro and Macro Mixing - Analysis, Simulation and Numerical Calculation*, chap. The Variance as Measured Variable for the Evaluation of a Mixing Process or for the Comparison of Mixtures and Mixers, pp. 2–16. Springer, Berlin, Heidelberg, 2010. doi:10.1007/978-3-642-04549-3.

[Tay21] Taylor. *Diffusion by Continuous Movements. Proceedings of the London Mathematical Society*, 20:196–212, 1921.

[Tay35] Taylor, G.I. *Statistical Theory of Turbulence. Proceedings of the Royal Society of London. Series A, Mathematical and Physical Sciences*, 151(873):421–444, 1935. ISSN 00804630.

[TKN19] T. Kalmár-Nagy, B.B. *An intriguing analogy of Kolmogorov's scaling law in a hierarchical mass–spring–damper model. Nonlinear Dyn*, 95:3193–3203, 2019. doi:https://doi.org/10.1007/s11071-018-04749-x.

[Van75] Van't Riet, K. and Smith, J.M. *The trailing vortex system produced by Rushton turbine agitators. Chemical Engineering Science*, 30(9):1093–1105, 1975. ISSN 0009-2509. doi:https://doi.org/10.1016/0009-2509(75)87012-6.

[War86] Warmoeskerken, M. *Gas-liquid dispersing characteristics of turbine agitators.* Ph.D. thesis, TU Delf, 1986.

[Web21] Weber, M. *WattsApp in Multiphase Systems by Gravity-Driven Buoyancy and Settling. Chemie Ingenieur Technik*, 93(1-2):108–116, 2021. doi:https://doi.org/10.1002/cite.202000103.

[Woz11] Woziwodzki, S. and Jędrzejczak, Ł. *Effect of Eccentricity on Laminar Mixing in Vessel Stirred by Double Turbine Impellers. Chemical Engineering Research and Design*, 89(11):2268–2278, 2011. ISSN 02638762. doi:10.1016/j.cherd.2011.04.004.

[Zeh02] Zehner, P. *Modelltechnik und Maßstabsübertragung*, chap. 16, pp. 405–423. John Wiley Sons, Ltd, 2002. ISBN 9783527603367. doi:https://doi.org/10.1002/3527603360.ch16.

[Zlo01] Zlokarnik, M. *Stirring: Theory and Practice.* Wiley-VCH, Weinheim ; New York, 2001. ISBN 978-3-527-29996-6. OCLC: ocm46693210.

A. Technical Data Sheets

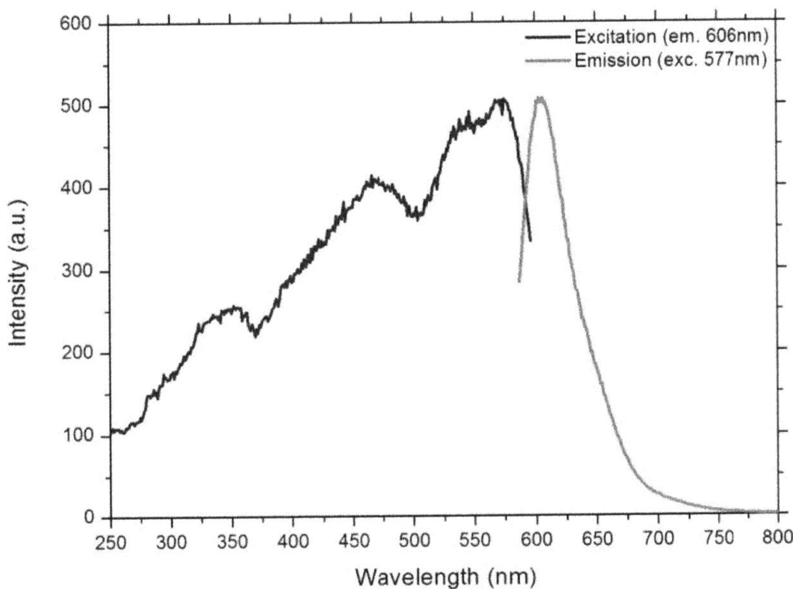

Figure A.1. Excitation and emission spectral response curves for fluorescent orange polyethylene microspheres 1.00g/cc - 0.01 to 1 mm

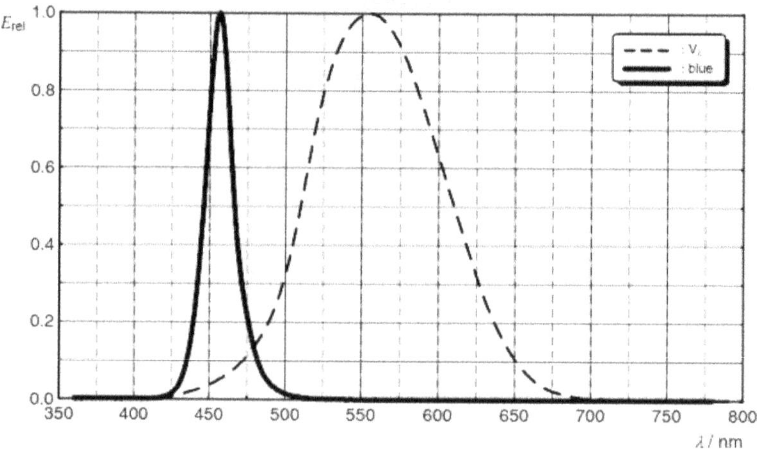

Figure A.2. Excitation and emission spectral response curves for fluorescent orange polyethylene microspheres 1.00g/cc - 0.01 to 1 mm

B. Additional Velocity and Residence Time Plots

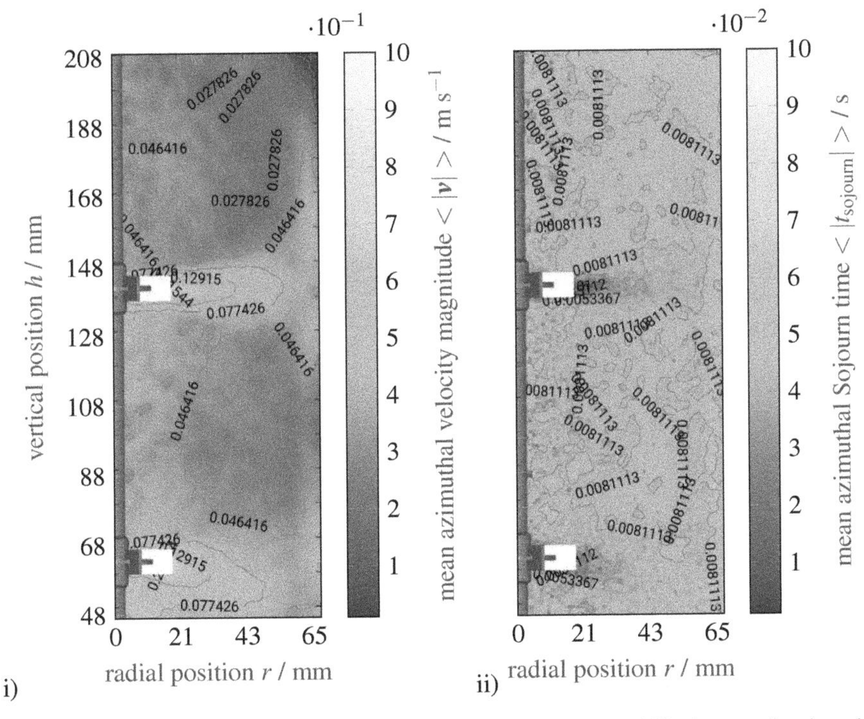

Figure B.1. i) Time and azimuthally averaged velocity map and ii) time and azimuthally averaged Sojourn time t_{sojourn}; double Rushton configuration; $n = 350\,\text{rpm}$; no aeration

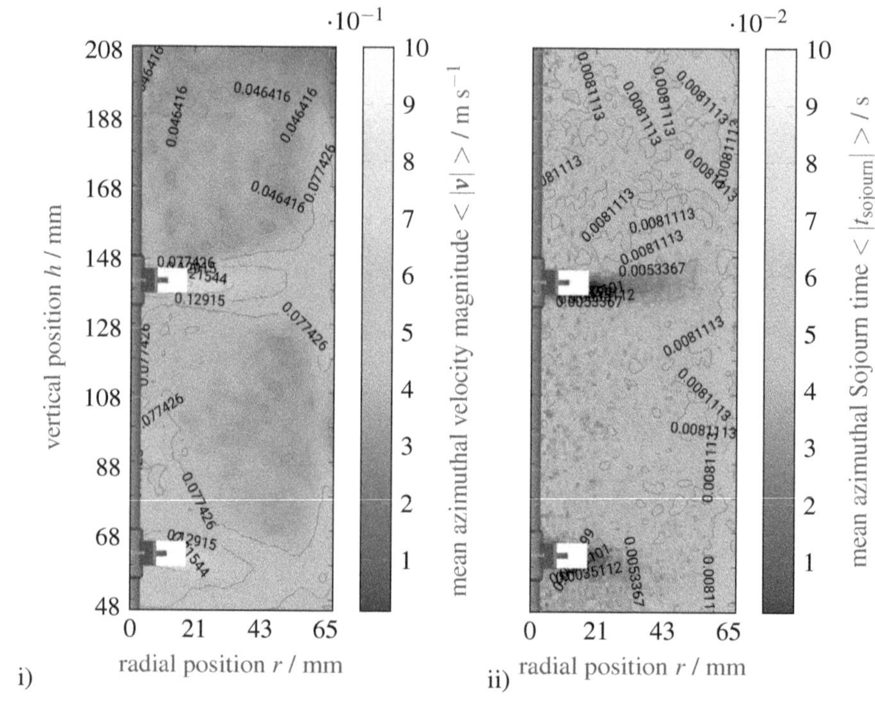

Figure B.2. i) Time and azimuthally averaged velocity map and ii) time and azimuthally averaged Sojourn time $t_{sojourn}$; double Rushton configuration; $n = 450$ rpm; no aeration

C. Overview of the Lagrangian Analysis Parameter and Results

Parameter	3x Elephant Ears			2x Rushton trubines		
stirrer diameter d / m	0.06	0.06	0.06	0.036	0.036	0.036
stirrer frequency n / s^{-1}	150	250	350	249	355	448
stirrer Reynolds number Re / -	9000	15000	21000	5378	7668	9676
spec. power input P/V / W m^{-3}	20.79	96.24	264.07	11.55	33.48	67.29
stirrer tip speed u_{tip} / m s^{-1}	0.4712	0.7854	1.0996	0.4694	0.6692	0.8445
Lagrangian autocorrelation time T_{L} / s	0.2700	0.1830	0.1100	0.3880	0.3543	0.2085
ballistic dispersion coefficients D_1 / m^2 s^{-2}	0.0081	0.0216	0.0450	0.0040	0.0080	0.0131
turbulent diffusive dispersion coefficients D_2 / m^2 s^{-1}	0.0049	0.0082	0.0122	0.0029	0.0041	0.0056
dimensionless ballistic dispersion coefficients D_1^* / -	0.0362	0.0351	0.0372	0.0198	0.0193	0.0197
dimensionless turbulent diffusive dispersion coefficients D_2^* / -	0.1720	0.1735	0.1852	0.1584	0.1546	0.1655
global mixing time Θ_{global} / s	8.19	5.76	5.31	11.37	7.72	6.38
dimensionless global mixing time $n \cdot \Theta_{\mathrm{global}}$ / -	20.48	24.00	30.98	47.19	45.65	47.64

Lebenslauf

Name	Fitschen
Vorname	Jürgen
Staatsangehörigkeit	Deutschland
Geburtsdatum	20.10.1990
Geburtsort	27404 Zeven

08.1997 - 07.2002	Grundschule Klostergang in Zeven
08.2002 - 07.2004	Orientierungsstufe in Zeven
08.2004 - 06.2011	St. Viti. Gymnasium in Zeven
11.2011 - 03.2015	Bachelorstudium Energie- und Umwelttechnik an der Technischen Universität Hamburg (TUHH), Abschluss: Bachelor of Science (B.Sc.)
04.2015 - 07.2017	Masterstudium Energie- und Umwelttechnik an der Technischen Universität Hamburg (TUHH), Abschluss: Master of Science (M.Sc.)
09.2017 - heute	Wissenschaftlicher Mitarbeiter am Institut für Mehrphasenströmungen an der Technischen Universität Hamburg (TUHH)

www.ingramcontent.com/pod-product-compliance
Ingram Content Group UK Ltd.
Pitfield, Milton Keynes, MK11 3LW, UK
UKHW061827190726
13853UKWH00009B/2470